Chemical Reaction Engineering

Nishith Verma

Chemical Reaction Engineering

Ane Books
Pvt. Ltd.

Nishith Verma
Department of Chemical Engineering
Department of Sustainable Energy
Engineering
Center for Environmental Science
and Engineering
Indian Institute of Technology Kanpur
Kanpur, Uttar Pradesh, India

ISBN 978-3-031-88690-4   ISBN 978-3-031-88691-1   (eBook)
https://doi.org/10.1007/978-3-031-88691-1

Jointly published with Ane Books Pvt. Ltd.
In addition to this printed edition, there is a local printed edition of this work available via Ane Books in South Asia (India, Pakistan, Sri Lanka, Bangladesh, Nepal and Bhutan) and Africa (all countries in the African subcontinent).
ISBN of the Co-Publisher's edition: 978-81-97888-69-4

This Springer imprint is published by the registered company Springer Nature Switzerland AG
The registered company address is: Gewerbestrasse 11, 6330 Cham, Switzerland

If disposing of this product, please recycle the paper.

Preface

ChE CRE follows *ChE Triad* published early this year by Ane Books Pvt. Ltd. *ChE Triad* is built upon my lecture notes of three courses that I have taught at the graduate level, namely: Transport Phenomena, Heterogeneous Chemical Reaction Engineering, and Numerical Methods. *ChE CRE* is built upon my lecture notes of "Chemical Reaction Engineering (CRE)" that I have taught at the undergraduate (UG) level. The book mainly deals with homogeneous chemical reactions. Few chapters, added towards the latter part of the book, deal with the basics of heterogeneous chemical reactions, as the detailed reactor design for heterogeneous chemical reactions is taught at the graduate level in most of the universities and institutes. Thus, *ChE CRE* covers the design of flow-reactors for homogeneous reactions, and replicates only the introductory topics on heterogeneous reactions, from the Heterogeneous Chemical Reaction Engineering part of *ChE Triad*. These topics are the shrinking core model for non-catalytic gas-solid reactions, Langmuir-Hinshelwood kinetics for catalytic reactions, and diffusion effects in porous catalysts.

Soon after *ChE Triad* was published, critiques pointed out that the publication should not have been termed a formal "book", rather a monogram book, i.e. collection of lectures. I cannot disagree. *ChE CRE* is also a monogram book, and the motivation for writing such a book is to present only those subjects or topics to the UG students which are absolutely necessary for designing a reactor or studying kinetics for homogenous chemical reactions. What I imply is that almost all text books prescribed for CRE at the UG level, including those of Levenspiel and Fogler contain the chapters on both homogenous as well as heterogeneous chemical reactions, and it is often left to the instructor to draw a line where to stop and what to cover from these books in a semester of around twenty five lectures of 90 min each. Considering that CRE is usually taught in the 3rd year of the UG program when the students are already exposed to stoichiometry, thermodynamics, fluid dynamics, unit operation, and a few numerical techniques, I have accordingly prepared *ChE CRE*.

ChE CRE comes with the audio lectures synchronized with the book chapters, and are freely downloadable from the web-link prescribed in the book. Students must use the audio lectures while going through the book chapters, as the description in the book is sketchy. I have tried explaining the topics as much as possible in the audio lectures.

I reproduce the following statements from the Preface of *ChE Triad*. Students should realize that a comprehensive learning comes only with rigor. Therefore, you should not use this book as a reading material; rather re-write the equations and solution steps shown in the book, in the same way as you take notes when an instructor is delivering lectures in an offline class.

To this end, I again close the preface with the same quote of Jiddu Krishnamurti, an Indian philosopher and a spiritual figure, which I mentioned in *ChE Triad*:

"There is no end to education. It is not that you read a book, pass an examination, and finish with education. The whole of life, from the moment you are born to the moment you die, is a process of learning".

Kanpur, India   Nishith Verma

2024

Competing Interests The author has no competing interests to declare that are relevant to the content of this manuscript.

Kinetics and Reactor Design

Syllabus

1. **Overview of chemical reaction engineering**:—definitions, nomenclature, notations, types of reaction, introduction to reactor design, types of reactors (batch versus flow, constant volume versus pressure)
2. **Kinetics of homogeneous reactions**: Rate expression, temperature and concentration dependence of rate, search for a reaction mechanism.
3. **Batch reactor**: Constant versus variable volume, interpretation of batch data for zeroth, 1st, 2nd, and nth order reactions, half-life, irreversible and reversible reactions, series and parallel reactions.
4. **Flow reactors (Mixed flow vs. plug flow reactors)**: Reactor design (size), performance equation, graphical representation, size comparison of reactors, multiple reactor systems (reactors in series and parallel), recycle reactor, autocatalytic reactions.
5. **Parallel reactions**: Selectivity, fractional yield, product distribution.
6. **Series reactions**: Product distribution and maximization of yield, qualitative description of series-parallel reactions.
7. **Non-isothermal reactions**: Thermodynamics of a reversible reaction, equilibrium conversion versus equilibrium constant, reactor design for exothermic and endothermic reaction, optimal progression for an exothermic reaction, adiabatic and non-adiabatic operation, multiplicity.
8. **Non-ideal flow reactors**: Segregated versus non-segregated flow, residense time distribution, tank in series model, dispersion model.
9. **Fluid-particle non-catalytic reactions**: Shrinking Core models for kinetics, diffusion (film) and ash layer controlled reactions, conversion versus time expressions, fluidized bed reactor
10. **Catalytic reactions**: Characterization; Langmuir-Hinshelwood models, diffusion and reaction in porous catalysts.

Text book: *Chemical Reaction Engineering* (3rd edition, Wiley) by Octave Levenspiel.

Reference books:

1. *Chemical Engineering Kinetics* (1981, McGraw) by J. M. Smith (recommended for introductory part, recycle reactors, non-ideal flow reactors, RTD)
2. *Elements of Chemical Reaction Engineering* (2nd edition, Prentice Hall) by Scott Fogler (recommended for chapters on multiplicity, non-catalytic solid-fluid reactions)

Contents

Chemical Processes

Introduction

Chemical transformation (birth or formation of a new species) from one species to another; may involve simultaneous physical processes such as evaporation, diffusion, condensation *etc.*

Examples:

$$C(s) + O_2(g) \rightarrow CO_2(g)$$
$$CaCO_3(s) + 2HCl(l) \rightarrow CaCl_2 + H_2O + CO_2$$

$H_2O + SS \rightarrow H_2O.SS?$ $\qquad\qquad\qquad H_2 + Br_2 \rightarrow 2HBr$

$(SS = \text{stainless steel})$ $\qquad\qquad\qquad N_2 + 3H_2 \rightarrow 2NH_3$

$CH_3COOH + NaCl \rightarrow CH_3COONa + HCl \qquad CO_2 + 3H_2 \rightarrow CH_3OH$

Chemical Reaction

Refers to the process indicating the type of species, along with conditions (T, P, *etc.*), involved in the chemical transformation. Reaction could be in the gaseous phase or liquid phase. Reactions could be isothermal, exothermic or endothermic, reversible or irreversible, single or multiple, parallel or in series.

Reactor Design

Size (volume) of the reactor?

(Detailed Design: T, P, rate, flow patterns, series, combination, type of reactor, recycle, yield, selectivity, *etc.*)

© The Author(s) 2025

N. Verma, *Chemical Reaction Engineering*,
https://doi.org/10.1007/978-3-031-88691-1_1

$$\text{Raw materials} \longrightarrow \boxed{} \to \overset{A \to B}{\boxed{Reactor}} \to \boxed{Post\text{-}treatment} \to Products$$

$$\swarrow \text{Pretreatment unit} \qquad\qquad \searrow \text{(washing, separation)}$$

Given: Reaction path, mechanism (overall), production rate (TPD, $\frac{Nm^3}{min}$, $\frac{m^3}{min}$), purity (pure B or unconverted A or some side products)

Homogeneous versus Heterogeneous reaction:

$1 - \phi$	$2 - \phi$ or multi-phase
Gas or Liquid	$(g/l + s,\, g + l + s,\, l + l)$
Catalytic or non-catalytic	Catalytic or non-catalytic
(microbes, enzymes in one phase)	
Examples:	
$H_2 + Br_2 \to 2HBr$	(1) $C + O_2 \to CO_2$
$CH_3COOH + NaCl(aq) \to CH_3COONa + HCl$	(2) $CaO + SO_2 \to CaSO_3$

Notes:

1. Simultaneous homogeneous and hetrogeneous reactions

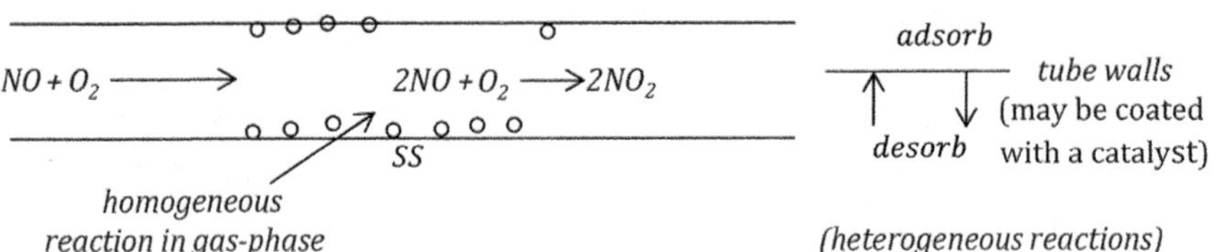

2. Catalytic reaction can be homogeneous or heterogeneous

$$\overset{monomer}{CH_3OH} \xrightarrow[80°]{\text{TEA (liq.)}} \overset{polymer}{n(CH_3OH)}; \qquad SO_2 + O_2 \xrightarrow[550°]{Pt\ (s)} SO_3$$

3. Biological conversion of organic into biomass in waste water (WW):

$$\underset{(suspended)}{Organics} + O_2 \xrightarrow{(microbes)} \underset{(biomass\ solid)}{Yeast} + CO_2$$

Considering micron-sized suspended solid particles in the reaction mixture, the reaction may be assumed to be homogeneous from design considerations.

Similarly, fermentors and fast fluidized bed reactors are often modeled/designed considering homogeneous reactions.

Rate of Reaction $(A \rightarrow B)$

$$-r_A \equiv \frac{moles\ of\ A\ converted}{time} \qquad \text{(not useful)}$$

$$\equiv \frac{moles\ of\ A\ consumed}{time \times (quantity)} \quad \text{-"useful"}$$

$$\text{which will make } (-r_A) \text{ intrinsic "meaningful"}$$

or *volume* / or *weight* / *surface area*

reaction mixture (liquid or solid or gas)

$$\text{moles/min-m}^3: \quad (-r_A) = -\frac{1}{V}\frac{dN_A}{dt} \Rightarrow (r_A \text{ is } - ve) \quad : gaseous\ or\ liquid\ phase\ (homogeneous)$$

(disapperance, + ve)

Rate of Reaction: $A + 2B \rightarrow C + D$

$$N_2 + 3H_2 \rightarrow 2NH_3$$

$$-r_{N_2} \quad -r_{H_2} \quad +r_{NH_3}$$

All three quantities are +ve; rates are always + ve.

Rate of Reaction (r)

$$r = \frac{1}{v_i}\frac{1}{\omega}\frac{dN_i}{dt}$$

+ ve

capacity (volume)

stoichiometric coefficient

$$v_i = \begin{cases} -ve & \text{for reactant} \\ +ve & \text{for products} \end{cases}$$

$\frac{dN_i}{dt}$ (time-gradient): $+ve$ or $-ve$, r will always be $+ve$.

$$-\frac{1}{1}\frac{1}{\omega}\frac{d(N_2)}{dt} = -\frac{1}{3}\frac{1}{\omega}\frac{d(H_2)}{dt} = \frac{1}{2}\frac{1}{\omega}\frac{d(NH_3)}{dt} = r$$

$$\left.\begin{array}{l} r_{H_2} = +3r_{N_2} \\ r_{NH_3} = -2r_{N_2} \end{array}\right\} r_{H_2},\ r_{N_2},\ r_{NH_3} \text{ will be } + \text{ or, } - \text{ but } r \text{ will always be } +ve$$

Similarly,

$$\text{For 1.}\quad A \to B;\qquad \begin{cases} r = +ve \\ r_A = -ve \\ r_B = +r \end{cases}$$

$$\text{2.}\quad A \to 2B;$$

$$r_A = -r,\quad r_B = 2r,\quad \Rightarrow r_B = -2r_A$$

$$\text{3.}\quad A + 2B \to C + 3D;$$

$$-\frac{1}{1}\frac{1}{\omega}\frac{d[A]}{dt} = -\frac{1}{2}\frac{1}{\omega}\frac{d[B]}{dt} = \frac{1}{1}\frac{1}{\omega}\frac{d[C]}{dt} = \frac{1}{3}\frac{1}{\omega}\frac{d[D]}{dt} = r$$

$$r_A = -r,\quad r_B = -2r,\quad r_c = r,\quad r_d = 3r$$

$$\boxed{r_A = \frac{r_B}{2} = -r_c = -\frac{r_d}{3}}$$

or

$$6r_A = 3r_B = -6r_c = -2r_d$$

In general and commonly;

$$aA + bB \to rR + sS \quad \left(\text{stoichiometric balance: } \sum v_A A = 0\right)$$

$$-r_A = -\frac{1}{V}\frac{dN_A}{dt} \quad \text{(amount of disappearance of } A/\text{volume-time; mol/m}^3\text{-s)}$$

where, $-r_A$ is $+ve$ and it is the rate of disappearance or rate of reaction of A.

$$-\frac{r_a}{a} = -\frac{r_b}{b} = \frac{r_R}{r} = \frac{r_s}{s} = \text{const} = r\ (+ve)$$

$$\searrow$$

$$\text{rate of reaction (unique of reaction)}$$

Moving on,

$$-r_A = f(C_A, C_B, T) \quad : experimental\ observation$$

$$concentration\ of\ A\ (moles/lt) \qquad temperature\ (energy\ required\ to\ break\ the\ chemical\ bond)$$

or,

$$= f(C_A)\phi(T) \qquad -\frac{1}{V}\frac{dN_A}{dt} = kf(C)$$

$$(if\ elementary) \qquad\qquad rate\ constant\ (T)$$

Order of Reaction: $A + B \rightarrow R$

$$-r_A = -\frac{1}{V}\frac{dN_A}{dt} = kC_A^\alpha C_B^\beta \text{ (rate equation can be simpler or complex)}$$

Order of reaction: $(\alpha + \beta)$; α-order with respect to A, β order with respect to B.

Note: Traditionally defined for elementary reactions.
Elementary Versus Non-elementary Reactions

(1)

$$\alpha A + \beta B \rightarrow \gamma R$$

If reaction (1) is elementary, that means 'α' # of A react (collide) with 'β' # of B to produce 'γ' # of R. As a consequence, we can write

$$-r_A = -\frac{1}{V}\frac{dN_A}{dt} = k[A]^\alpha [B]^\beta$$

It is, therefore, clear that α, β, γ must +ve integer, and $\alpha + \beta$ and $\gamma \leq 3$ (3-body collision). Also, note that no reaction is truely irreversible. Therefore, all arguments for elementary or non-elementary reactions hold good for reverse reaction also.

$$\gamma R \rightarrow \alpha A + \beta B$$

Mathematically, if it is a non-elementary reaction then $(-r_A)$ could assume any form (simple or complex)

$$-r_A = \frac{k_1 C_A}{k_2 + k_1 C_A C_B^2}; \quad k_1 C_A; \quad k_2 C_A^2; \quad k_1 C_A^{0.5} C_B, \quad etc.$$

There is no order, overall or with respect to singular species in the previous 1st rate expression. If

$$-r_A = k_1 C_A^{\alpha-1} \quad \text{or} \quad -r_A = k_1 C_A C_B^{\beta-1},$$

physically, what does it mean? A non-elementary reaction is actually a sequence of elementary reactions which occur in true sense.

(2) For 'irreversible' reactions:

$$A \rightarrow B; \quad -r_A = kC_A$$
$$2A \rightarrow B; \quad -r_A = kC_A^2$$
$$2A + B \rightarrow C; \quad -r_A = kC_A^2 C_B$$
$$2A \rightarrow 2B; \quad -r_A = kC_A^2$$

(3)

$$2A \rightarrow 2B; \quad (-r_A) = kC_A \Rightarrow \text{non-elementary reaction}$$

(4) For a non-elementary reaction; 'order' could be fraction, negative.

$$2A \rightarrow B; -r_A = k_1 C_A^{\frac{1}{2}}; -r_A = C_A^{-\frac{1}{2}}$$

(a) Mechanism decides whether a reaction is elementary or non-elementary.

(b) Stoichiometry decides how $(r_B, etc.)$ are related with r_A

$$SO_2 + 2O_2 \rightarrow 2SO_3 \quad - \quad \text{(a)}$$
$$2SO_2 + 4O_2 \rightarrow 4SO_3 \quad - \quad \text{(b)}$$
$$\frac{1}{2}SO_2 + O_2 \rightarrow SO_3 \quad - \quad \text{(c)}$$

In all (a), (b) and (c)

$$-r_{SO_2} = -\frac{r_{O_2}}{2} = \frac{r_{SO_3}}{2} = r, \quad -r_{SO_2} = \frac{k_1 [SO_2]}{k_2 + k_3 [O_2]^{\frac{1}{2}}}$$

(The reaction is non-elementary).

Non-elementary Reaction

Example:

$$H_2 + Br_2 \rightleftharpoons 2HBr$$

Experimental data indicate

$$-r_{H_2} \neq k C_{H_2} C_{Br_2}, \quad r_{HBr} = -2r_{H_2}$$

In fact,

$$-r_{HBr} = \frac{k_1 [H_2] \left[Br_2^{\frac{1}{2}} \right]}{k_2 + [HBr]/[Br_2]} \quad \text{(see later)}$$

The rate expression clearly indicates that the reaction path is something else; it consists of several intermediate steps, which are rather *elementary*.

Caution:

(a)

$$Br_2 \underset{k_2}{\overset{k_1}{\rightleftharpoons}} 2Br$$

k_1, k_2 are rate constants.

$$-\frac{d[Br_2]}{dt} = -r_{Br_2} = k_1[Br_2] - \frac{k_2[Br]^2}{2}$$

$$\left(\frac{r_{Br_2}}{-1} = \frac{r_{Br}}{2}; \ \frac{r_{Br}}{-2} = \frac{r_{Br_2}}{1} \right)$$

© The Author(s) 2025

N. Verma, *Chemical Reaction Engineering*,

https://doi.org/10.1007/978-3-031-88691-1_2

or,

$$Br_2 \xrightarrow{k_1} 2Br^\bullet$$

$$2Br^\bullet \xrightarrow{k_2} Br_2$$

(k_1 and k_2 must be specified with respect to reactant).
Therefore,

$$r_{Br} = 2k_1[Br_2] - k_2[Br]^2$$
$$= 2\left[-r_{Br_2}\right]: \quad \text{overall.}$$

(b)

$$B + 2D \xrightarrow{k_1} 3T \quad \text{(assume irreversible)}$$

where, k_1 is the rate constant.
If elementary, $-r_B = k_1 C_B C_D^2$? and $-r_D = k_1 C_B C_D^2$? No.

$$(-r_B) = -\left(\frac{r_D}{2}\right)$$

Usually, the rate content refers to the first species.

$$\left.\begin{array}{l} -r_B = k_1 C_B C_D^2 \\ -r_D = 2k_1 C_B C_D^2 \\ +r_T = 3k_1 C_B C_D^2 \end{array}\right\}$$

One has to be explicit which component it is referring to.
For example,

$$\text{if} \quad -r_B = 0.23 C_B C_D^2,$$
$$-r_D = 0.46 C_B C_D^2$$
$$\left(r = \text{ rate of reaction } = 0.23 C_B C_D^2\right)$$

Rate constant

$$r_B = -r, \, r_D = -2r, \, r_T = +3r$$

Mechanism: (The step involves short-lived species)

1. Free radicals: $CH_3^\bullet$, $Br^\bullet$, $H^\bullet$
2. Transition complex (intermediate/unstable, A^*)

$$A + A \rightarrow A^* + A$$
$$A^* + R \rightarrow AR^*$$
$$AR^* \rightarrow S$$

$\Longrightarrow$ Propose a hypothesis based on experimental observations, and plausible theories. Some thumb's rules can verify the experimental data. 'It means two different mechanisms can give rise to one rate of reaction'. Also today's theory may be proven wrong tomorrow.

$$N_2 + 3H_2 \rightarrow 2NH_3$$

$(-r_A)$ can assume an identical expression by two different pathways.

$$r_i = \frac{f_i\ (\text{temperature})}{(k(T),\ T)} \cdot \frac{f_2\ (\text{composition})}{\left(C_A C_B,\ \text{or}\ C_A^2,\ \text{etc.}\right)} \left.\right\} \begin{array}{l}\text{for elementary}\\ \text{reaction}\end{array}$$

1. **Arrhenius Law:**

$$k = k_o \exp\left(-\frac{E}{RT}\right) \ for\ an\ elementary\ reaction$$

where the arrow up points to $Activation\ energy\ \left(\dfrac{cal}{mole}\right)$ and the arrow down points to $frequency\ factor,\ pre\text{-}expoential\ factor$.

$R =$ Universal gas constant and $T =$ Reaction temperature.

Assuming

$$A \underset{k_2}{\overset{k_1}{\rightleftharpoons}} B \ to\ be\ an\ elementary\ reaction$$

$$k_1 = k_{10} \exp\left(-\frac{E_1}{RT}\right) \text{ and } k_2 = k_{20} \exp\left(-\frac{E_2}{RT}\right)$$

This implies that there is an 'energy barrier' for both forward and backward reactions.

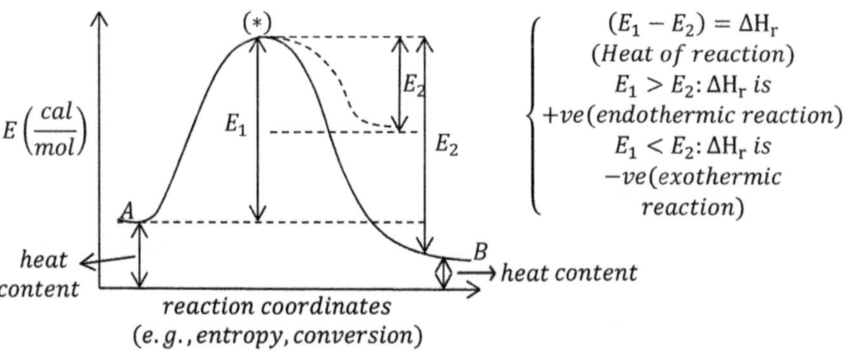

$\Longrightarrow$ (catalyst indicates an alternate path in which the activation energy is lower; heat contents (thermodynamic properties) of 'A' and 'B' remain the same)

Thermodynamics:

Van't Hoff: $A \underset{k_2}{\overset{k_1}{\rightleftharpoons}} B$

$$K = \frac{k_1}{k_2} \text{ (equilibrium constant)}$$

$$\frac{d(\ln K)}{dT} = \frac{\Delta H_r}{RT^2} \text{ (At equilibrium } r = r_f - r_b = 0)$$

or

$$\frac{d \ln k_1}{dT} - \frac{d \ln k_2}{dT} = \frac{E_1}{RT^2} - \frac{E_2}{RT^2}$$

Suggestion:

$$\frac{d \ln k_1}{dT} = \frac{E_1}{RT^2} \Rightarrow \ln k_1 = -\frac{E_1}{RT} + constant$$

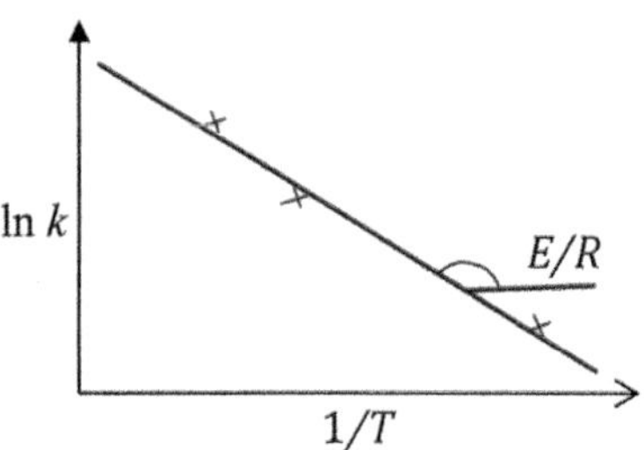

$$k = k_o \exp\left(-\frac{E}{RT}\right) \Longrightarrow \text{Arrhenius's law}$$

For **non-elementary reaction**: There is no activation energy barrier diagram. One can define an apparent activation energy, which could be negative also.

(2) **Collision Theory:**

$$k \propto T^{\frac{1}{2}} \exp\left(-\frac{E}{RT}\right)$$

(3) **Transition-state Theory:**

$$A + B \underset{k_2}{\overset{k_1}{\rightleftharpoons}} AB^* \xrightarrow{k_3} AB \quad (AB^* \equiv \text{short-lived transient species})$$

$$k \propto T \exp\left(-\frac{E}{RT}\right)$$

(1) Equilibrium between: AB^* and $A + B$

(2) $AB^* \to AB$

Difference between

Collision Theory	Transition-State Theory
Intermediate breakage does not influence the rate	Rate of formation is in equilibrium & there is a rate in the next step

In general:

$$k \propto T^m \exp\left(-\frac{E}{RT}\right)$$

Arrhenius law is the most widely used

$$\boxed{k \propto k_o \exp\left(-\frac{E}{RT}\right) \quad (m = 0)}$$

Activation Energy and temperature dependency

$$A \to B; \quad E \neq f(T)$$

$$k = k_o \exp\left(-\frac{E}{RT}\right)$$

(4) Let us calculate temperature rise needed to double the rate of reaction. That is, $\Delta T = T_2 - T_1$?

If $k_2 = 2k_1$ or

$$2 = \exp\left(-\frac{E}{R}\left(\frac{1}{T_2} - \frac{1}{T_1}\right)\right)$$

$$\longleftarrow E \text{ (cal/mol)} \longrightarrow$$

$T_{(\circ C)}$	40	160	280	400	
0	11	2.7	1.5	1.1	
400	65	16	9.3	6.5	$\to \Delta T^\circ C$
1000	233	58	33	23	
2000	744	185	106	74	

(1) Reactions with high activation energies are very temperature sensitive; reactions with low activation energies are relatively temperature insensitive - Row wise.

(2) A given reaction (E is fixed) is much more temperature sensitive at low temperature than at high temperature: column wise.

(1) can be explained for a reversible exothermic or endothermic reaction

$$A \underset{k_2}{\overset{k_1}{\rightleftharpoons}} B, \quad \text{if exothermic, } E_2 > E_1, \Delta H_r \text{ is } -\text{ve}$$

Increasing temperature will cause r_f, r_b both to increase

$$k = k_o \exp\left(\frac{-E_1 \text{ or } E_2}{RT}\right)$$

But r_b increases at higher rate than r_f and eventually another equilibrium will be reached at lower conversion.

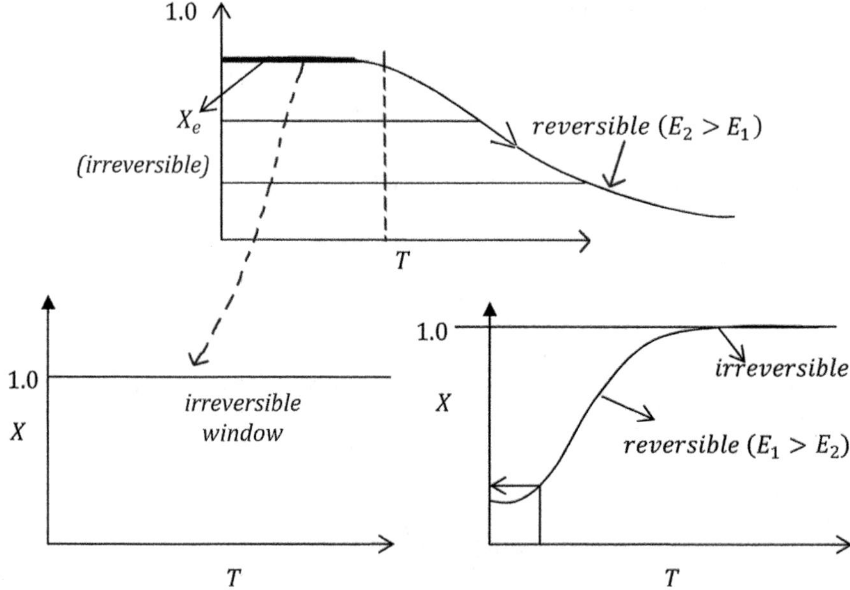

If endothermic, $E_1 > E_2$, ΔH_r is +ve. Increasing temperature will cause r_f, r_b ↑, but r_f will increase at higher rate than r_b, the net rate will increase; and equilibrium conversion will increase; may reach 1.0.

Searching for a Mechanism (Basis)

(1) Stoichiometry can not tell us whether we have a single reaction pathway or not

$$A \rightarrow 1.45R + 0.85s \rightarrow \quad \text{non-elementary.}$$

(2) No elementary reaction with molecularity greater than '3' has been observed to date:

$$N_2 + 3H_2 \rightarrow 2NH_3 \rightarrow \quad \text{non-elementary.}$$

(3)

$$2NH_3 \underset{2}{\overset{1}{\rightleftharpoons}} N_2 + 3H_2$$

(2) is not acceptable, so (1) cannot be acceptable. The reaction is non-elementary.

(4) Simultaneous rupture of bond, molecular synthesis, or splitting are likely to occur one at a time, each being an elementary step in the mechanism.

$$2NH_3 \rightleftharpoons (NH_3)_2^* \rightleftharpoons N_2 + 3H_2 \quad \text{is very unlikely.}$$

(5) Rate of reaction for a reactive species (radical, transient complex) can be assumed to be 'faster' in comparison with that of competitive reactions - **Quasi-steady state (Pseudo-steady state)**

$$\left(\frac{d\,(R^* \cdot)}{dt} = 0, \quad \text{with respect to} \quad \frac{d(\text{major species})}{dt} \right)$$

© The Author(s) 2025

N. Verma, *Chemical Reaction Engineering,*
https://doi.org/10.1007/978-3-031-88691-1_3

Examples:

(1)

$$H_2 + Br_2 \rightleftharpoons 2HBr \quad \text{(overall stoichiometric reaction)}$$

$$r = -\frac{d(H_2)}{dt} = -\frac{d(Br_2)}{dt} = \frac{1}{2}\frac{d(HBr)}{dt}$$

(a) Mechanism:

$$Br_2 \underset{k_{-1}}{\overset{k_1}{\rightleftharpoons}} 2Br^{\bullet} \quad \text{(Initiation/Termination)}$$

$$\left.\begin{array}{c} \bullet Br + H_2 \underset{k_{-2}}{\overset{k_2}{\rightleftharpoons}} HBr + H^{\bullet} \\[2ex] \bullet H + Br_2 \underset{k_{-3}}{\overset{k_3}{\rightleftharpoons}} HBr + Br^{\bullet} \end{array}\right\} Propagation$$

(b) Assumptions $k_{-3} \ll k_3$,

$$\bullet H + Br_2 \gg HBr + Br^{\bullet}$$

(c)

$$-\frac{dBr_2}{dt} = k_1(Br_2) - \frac{k_{-1}(Br^{\bullet})^2}{2} + k_3(\bullet H)(Br_2)$$

(d) Quasi-steady state assumptions for $(\bullet H)$ and $(Br^{\bullet})$ are:

$$\frac{dH^{\bullet}}{dt} = k_2(Br^{\bullet})(H_2) - k_2(HBr)(\bullet H) - k_3(\bullet H)(Br_2) = 0$$

$$\begin{aligned} \frac{d(Br^{\bullet})}{dt} &= 2k_1(Br_2) - k_{-1}(Br^{\bullet})^2 - k_2(Br^{\bullet})(H_2) \\ &\quad + k_3(Br_2)(H^{\bullet}) + k_{-2}(HBr)(\bullet H) \\ &= 0 \end{aligned}$$

(e) Substituting,

$$r = \frac{k_3 k_2 \sqrt{\frac{k_1}{k_{-1}}}(Br_2)^{\frac{3}{2}}(H_2)}{k_{-2}(HBr) + k_3(Br_2)}$$

–Temperature dependence and concentration dependence terms cannot be separated. What is E?? Not defined for a non-elementary reaction.

Some special cases:

$$r = \frac{k_3 k_2 \sqrt{\frac{k_1}{k_{-1}}} (Br_2)^{\frac{1}{2}} (H_2)}{k_3} \quad : \text{ begining of reaction}$$

$$= k_2 \underbrace{\sqrt{\frac{k_1}{k_{-1}}}}_{f(k)} \underbrace{(Br_2)^{\frac{1}{2}} (H_2)}_{f(C)}$$

$$= \underbrace{\frac{k_3 k_2}{k_{-2}} \sqrt{\frac{k_1}{k_{-1}}}}_{f(k)} \underbrace{\frac{(Br_2)^{\frac{3}{2}} (H_2)}{(HBr)}}_{f(C)} \quad : \text{ later stage of reaction}$$

Order of reaction = fraction for a non-elementary reaction?

$$r = k_{obs} f(C) \Rightarrow k_{obs} = k_2 \sqrt{\frac{k_1}{k_{-1}}} \quad \text{(begining stage of reaction)}$$

$$\Rightarrow E_{obs} = E_2 + \frac{E_1 - E_{-1}}{2} = +\text{ve or} -\text{ve}$$

All possibilities for a non-elementary reactions slopes (E_{obs} +ve or $-$ve):

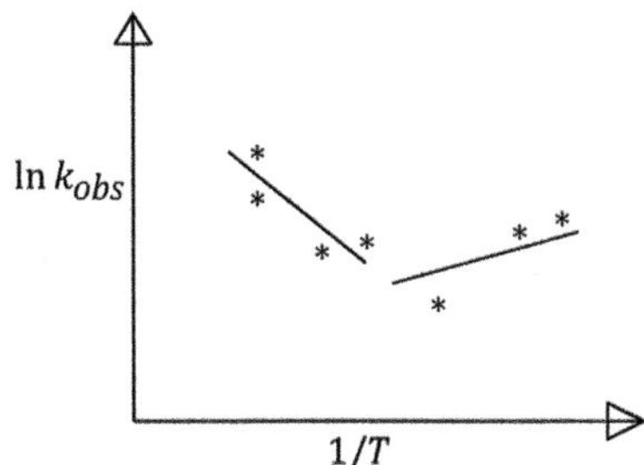

E_{obs} does not have any physical meaning; no energy barrier or energy diagram; just a mathematical no. E_1, E_2, E_{-1} are always +ve.

(2)

$$CO + H_2O \xrightarrow[\text{Fe Oxide}]{T - 450^\circ C} CO_2 + H_2 \quad \text{(water-gas shift reaction)}$$

(Note that we are discussing a heterogeneous reaction)

Active sites on a catalyst

$\downarrow$

$\longrightarrow\!\times\!\!-\!\!\times\!\longrightarrow\!\times\!\!-\!\!\times\!\longrightarrow\!\times\!\!-\!\!\times\longrightarrow$

'X' : electronically active

$$X + H_2O \xrightarrow{k_1} XO + H_2 \quad \text{(elementary)}$$

$$XO + CO \xrightarrow{k_2} CO_2 + X \quad \text{(elementary)}$$

(1)

$$r = -\frac{dCO}{dt} = -\frac{dH_2O}{dt}$$

(2)

$$\left.\begin{aligned} +\frac{d[X]}{dt} &= k_2[XO][CO] - k_1[X][H_2O] = 0 \\ \frac{d[XO]}{dt} &= k_1[X][H_2O] - k_2[XO][CO] = 0 \end{aligned}\right\} \text{PSSA/QSSA}$$

$$[XO] = \frac{k_1[X][H_2O]}{k_2[CO]}$$

$X + XO = A$; at any time, site is either X or XO. 'A' is the property of the catalyst, *viz.* number of active sites per amount of the catalyst.

$$r = \frac{Ak_1k_2[CO][H_2O]}{k_1[H_2O] + k_2[CO]} \neq f(T)\phi(C)$$

– Reaction-order is not defined here *e.g.*,

$$\left(\text{for } r = k_1(c_1)^\alpha \cdot (c_2)^\beta \ldots, \text{order} = (\alpha + \beta + \gamma)\right)$$

– Activation energy is not defined here. However if $k_1 \gg k_2$ or $k_2 \gg k_1$, then order or E can be defined as apparent order or E_{obs}.
If $k_1 \gg k_2$

$$r = Ak_2[CO]$$

if $k_1 \ll k_2$

$$r = Ak_1[H_2O]$$

(In many homogeneous reactions, catalyzed by enzymes, 'X' is denoted as E and A with E_0)

(3) Decomposition of N_2O_5:

$$2\,N_2O_5 \rightarrow 4NO_2 + O_2$$

The observate rate is 1st -order. Propose a mechanism

(i)

$$N_2O_5 \underset{k_2}{\overset{k_1}{\rightleftharpoons}} NO_2 + NO_3$$

(ii)

$$NO_3 \xrightarrow{k_3} NO + O_2$$

Intermediates are NO and NO_3

(iii)

$$NO + NO_3 \xrightarrow{k_4} 2NO_2$$

(a)

$$r = -\frac{1}{2}\frac{d\,[N_2O_5]}{dt}$$

(b)

$$-\frac{d\,[N_2O_5]}{dt} = k_1\,[N_2O_5] - k_2\,[NO_2]\,[NO_3]$$

(c)

$$\frac{d\,[NO_3]}{dt} = \frac{d[NO]}{dt} = 0: \quad (\text{PSSA/QSSA})$$

$$\frac{d\,[NO_3]}{dt} = k_1\,[\,N_2O_5] - k_2\,[NO_2]\,[NO_3] - k_3\,[NO_3]$$

$$- k_4[NO]\,[NO_3] = 0 \quad (\text{PSSA})$$

$$\frac{d[NO]}{dt} = k_3\,[NO_3] - k_4[NO]\,[NO_3] = 0 \quad (\text{PSSA})$$

$$[NO] = \frac{k_3}{k_4}$$

$$[NO_3] = \frac{k_1[N_2O_5]}{(2k_3+k_2)[NO_2]} \quad\rightleftarrows\quad \textit{measurable quantities}$$

(d)

$$r = -\frac{1}{2}\frac{d\,[N_2O_5]}{dt} = \frac{k_1k_3}{2k_3 + k_2}\,[N_2O_5]: \quad \text{1st-order}$$

(NO, NO_3 are highly reactive species: PSSA/QSSA holds good)

(4)

$$2A + B \rightarrow A_2B : r_{A_2B} = k[A][B]: \quad \text{2nd-order observed reaction}$$

(a)

$$A + B \xrightarrow{k_1} AB^*$$

$$AB^* + A \underset{k_3}{\overset{k_2}{\rightleftharpoons}} A_2B$$

$$-\frac{dC^*_{AB}}{dt} = -r^*_{AB} = k_1[A][B] - k_2[AB^*][A] + k_3[A_2B] = 0$$

$$[AB^*] = \frac{k_1[A][B] + k_3[A_2B]}{k_2[A]}$$

$$\frac{dC^*_{A2B}}{dt} = r_{A_2B} = k_2[AB^*][A] - k_3[A_2B] = k_1[A][B] \quad \text{(2nd order)}$$

(Two different mechanisms can show the same overall reaction rate! See next)

(b)

$$A + B \underset{k_2}{\overset{k_1}{\rightleftharpoons}} AB^*$$

$$AB^* + A \xrightarrow{k_3} A_2B$$

$$r^*_{AB} = k_1[A][B] - k_2[AB^*] - k_3[A][AB^*] = 0 \quad (PSSA)$$

$$[AB^*] = \frac{k_1[A][B]}{k_2 + k_3[A]}$$

$$r_{A_2B} = \frac{k_3[A]k_1[A][B]}{k_2 + k_3[A]} = k_1[A][B] \quad \left(\text{if } [A]\frac{k_3}{k_2} \gg 1\right)$$

$$\text{(2nd order)}$$

Constant-Volume Batch Reactor (Isothermal)

4

Batch versus Flow:	Reactor types
Liquid versus Gas:	Fluid Property
Constant P versus Constant V:	Operation

Note:

$$A \rightarrow 2B : \Delta n = 2 - 1 = 1$$

$$A \rightarrow B : \quad \Delta n = 0$$

$\left.\right\}$ gas or liquid

$$-r_A = -\frac{1}{V}\frac{dN_A}{dt} = -\frac{d}{dt}\left(\frac{N_A}{V}\right) = -\frac{dC_A}{dt} \begin{cases} \text{isothermal or non-isothermal} \\ \text{constant } V \end{cases}$$

Ideal gas:

$$r_A = \frac{1}{RT}\frac{dp_A}{dt} \text{ (isothermal)}$$

Thus, gas-concentration is related to partial pressure.

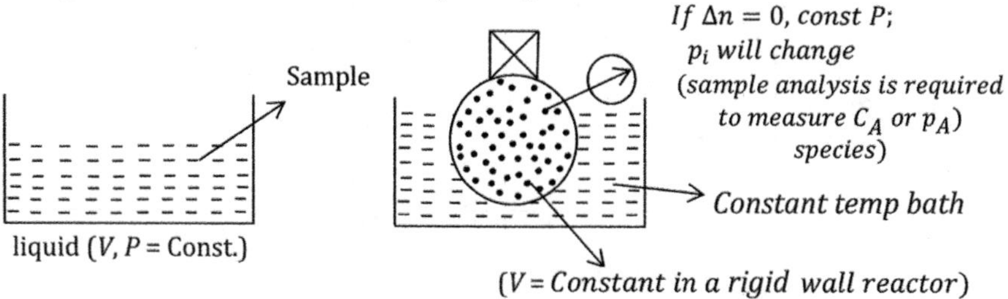

© The Author(s) 2025

N. Verma, *Chemical Reaction Engineering*,

https://doi.org/10.1007/978-3-031-88691-1_4

Fractional Conversion:

$$X_A = \frac{N_{AO} - N_A}{N_{AO}} = 1 - \frac{N_A}{N_{AO}} = \left(1 - \frac{C_A}{C_{AO}}\right) = \frac{C_{AO} - C_A}{C_{AO}}$$

(Batch system)

$$dX_A N_{AO} = -dN_A$$

Const. V.

$$C_A = C_{AO}(1 - X_A)$$

for const. V only

$$-dC_A = C_{AO}dX_A$$

X_A should be used for single stoichiometry.

Determining rate constants:

1. Integral Method of Analysis of data
2. Differential Method of Analysis of data

(a) Irreversible unimolecular 1st order: $A \rightarrow$ Products $(B + 2C)$

$$\left.\begin{array}{l} C_A = C_{AO} - C_{AO}X_A \\ C_B = C_{BO} + C_{AO}X_A \\ C_C = C_{CO} + 2C_{AO}X_A \end{array}\right\} \quad -r_A = -\frac{dC_A}{dt} = kC_A \Rightarrow C_A = C_{AO}e^{-kt}$$

or

$$\ln \frac{C_A}{C_{AO}} = -kt \text{ or } \ln(1 - X_A) = -kt$$

Batch experimental data:

t	C_A
0	C_{AO}
⋮	⋮
⋮	⋮

Plot:

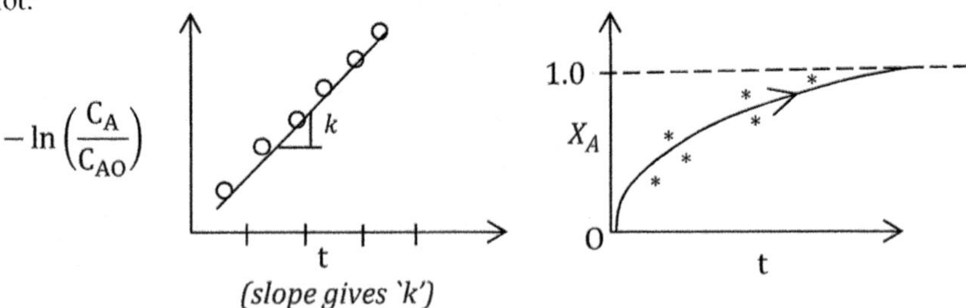

(b) $A + B \rightarrow$ Products

$$-r_A = -\frac{dC_A}{dt} = -\frac{dC_B}{dt} = kC_A C_B \quad \begin{pmatrix} C_A = C_{AO}(1 - X_A) \\ C_B = C_{BO} - C_{AO} X_A \\ C_B = C_{BO} - C_{BO} X_B \\ C_{AO} X_A = C_{BO} X_B \end{pmatrix}$$

$$+ C_{AO}\frac{dX_A}{dt} = kC_{AO}(1 - X_A)\, C_{AO}(M - X_A) \quad \left(M = \frac{C_{BO}}{C_{AO}}\right)$$

or

$$+ \frac{dX_A}{dt} = k(1 - X_A)\, C_{AO}(M - X_A)$$

(c)

$$\int_0^{X_A} \frac{dX_A}{(1 - X_A)(M - X_A)} = C_{AO}k \int_0^t dt$$

$$\ln \frac{M - X_A}{M(1 - X_A)} = C_{AO}(M - 1)kt$$

$$= (C_{BO} - C_{AO})kt,\ M \neq 1,\ C_{AO} \neq C_{BO}$$

or

$$\left(\ln \frac{1 - X_B}{1 - X_A} = \ln \frac{C_B}{MC_A} = \ln \frac{C_B C_{AO}}{C_{BO} C_A} \right)$$

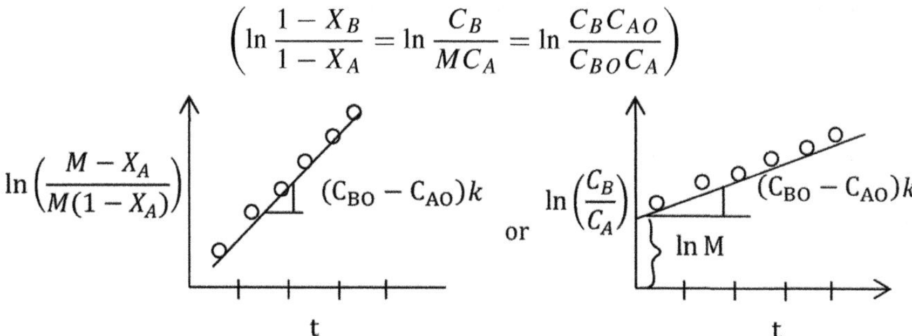

or

If $C_{BO} \gg C_{AO}$, i.e.,

$$100\frac{\text{mol}}{\text{lt}}\, B, \quad 1\frac{\text{mol}}{\text{litre}}\, A \rightarrow M = 100$$

$$-r_A = kC_A C_B = k'C_A \quad (k' = kC_{BO})$$

(C_B may be assumed to be constant at C_{BO}). Equation reduces to that in (a)

$$\ln \frac{1 - \frac{X_A}{M}}{1 - X_A} = C_{AO} M k t \quad (M \gg X_A \text{ (fraction } < 1))$$

$$\left(\ln \frac{1}{1 - X_A} = C_{BO} k t = k't \right) \quad \text{as long as } C_A \gg C_B \text{ or } M \gg 1$$

$$\ln (1 - X_A) = -k't$$

(d) $M = 1$ or $2A \to B$, $-r_A = kC_A^2$

$$-r_A = -\frac{dC_A}{dt} = kC_{AO}^2 (1 - X_A)^2$$

$$\frac{1}{C_A} - \frac{1}{C_{AO}} = kt = \frac{1}{C_{AO}} \frac{X_A}{1 - X_A}$$

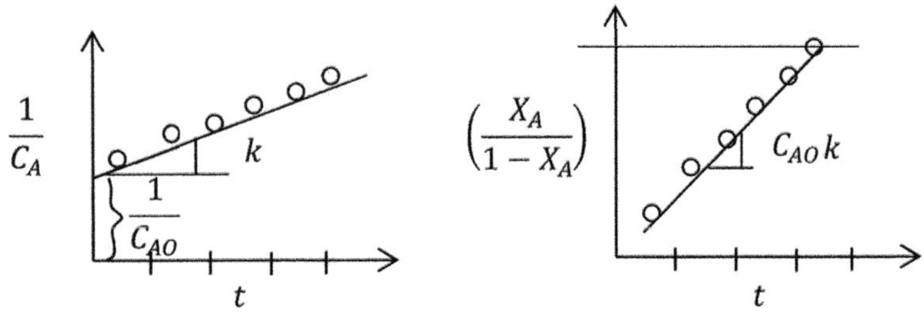

Caution: $A + 2B \to$ Product; $-r_A = kC_A C_B$ (non-elementary reaction)

$$C_A = C_{AO} (1 - X_A)$$

$$C_B = C_{BO} - 2C_{AO} X_A$$

(Note $C_B \neq C_{Bo} - C_{Ao} X_A$) (do fresh calculations).

(e) Zero th order: $A \to$ Product

$$-r_A = -\frac{dC_A}{dt} = k \text{ (independent of concentration)}$$

(C_A remains constant for long time; A is in excess)

$$\equiv \text{ model (non-elementary reaction)}$$

$$(C_{AO} - C_A = kt = C_{AO} X_A), \ t < \frac{C_{AO}}{K}$$

$$C_A = 0, \ t \geq \frac{C_{AO}}{K}$$

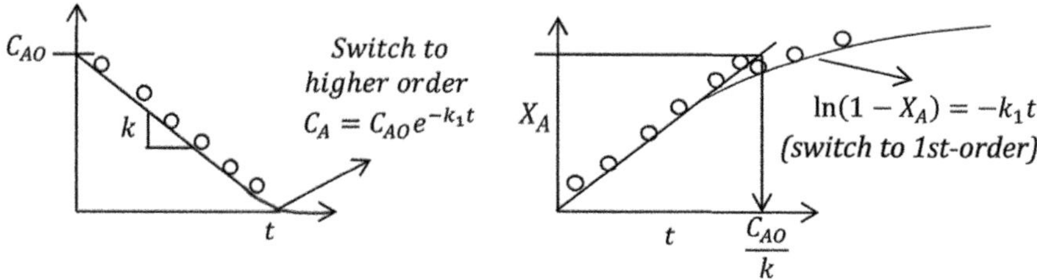

(No reaction will remain zeroth order for ever; it switches to 1st-order at some time when the concentration of A decreases to some extent)

(f) Irreversible reactions in series

$$\left. \begin{array}{l} t = 0,\, C_A = C_{AO} \\ C_{RO} = C_{SO} = 0 \end{array} \right] \quad A \xrightarrow{k_1} R \xrightarrow{k_2} S$$

Intermediate? 'R' may be considered to be an intermediate. It can also be a desired product.

(1)

$$-r_A = -\frac{dC_A}{dt} = +k_1 C_A \Rightarrow C_A = C_{AO}e^{-k_1 t} \tag{1}$$

    (Independent of any successive reaction)

(2)

$$r_R = \frac{dC_R}{dt} = k_1 C_A - k_2 C_R \Longrightarrow \frac{dC_R}{dt} = k_1 C_{AO}e^{-k_1 t} - k_2 C_R$$

(3)

$$r_s = \frac{dC_s}{dt} = k_2 C_R \quad \left(\frac{dy}{dx} + P(y) = Q(t) \right)$$

(4)

$$C_R = C_{AO}k_1 \left(\frac{e^{-k_1 t}}{(k_2 - k_1)} + \frac{e^{-k_2 t}}{(k_1 - k_2)} \right), \quad k_1 \neq k_2 \tag{2}$$

From stichometry

$$\left. \begin{array}{l} C_{AO} - C_A = C_R + C_S \\ C_{AO} = C_A + C_R + C_S \\ C_S = C_{Ao} - C_A - C_R \end{array} \right\}$$

(Add (1), (2), (3):

$$\frac{d}{dt}(C_A + C_R + C_S) = 0; \quad C_A + C_R + C_S = C_{Ao})$$

$$C_S = C_{AO}\left(1 + \frac{k_2}{k_1 - k_2}e^{-k_1 t} + \frac{k_1}{k_2 - k_1}e^{-k_2 t}\right) \qquad (3)$$

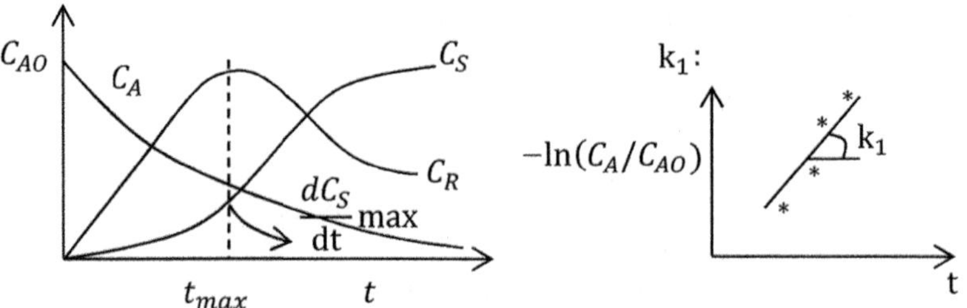

Differentiate (2) $dC_R/dt = 0$, time when C_R is maximum.

$$t_{max} = \frac{\ln\left(\frac{k_2}{k_1}\right)}{(k_2 - k_1)} = \frac{1}{\text{log-mean }(k_2\&k_1)}$$

$$C_{R,max} = C_{AO}\left(\frac{k_1}{k_2}\right)^{\frac{k_2}{(k_2-k_1)}} \quad (k_2 \text{ can also be determined from here})$$

Data:

t	$\cdots$	$\cdots$	$\cdots$	t_{max}	$\cdots$
C_R	0	2	3	$C_{R,max}$	4

$$A \xrightarrow{k_1} R \xrightarrow{k_2} S \quad \text{(simplify the previous solution)}$$

Two cases:

(1) $k_1 \gg k_2$
$C_A = C_{AO}e^{-k_1 t}$
$C_R = C_{AO}(e^{-k_2 t} - e^{-k_1 t})$

$C_S = C_{AO}(1 - e^{-k_2 t})$

(2) $k_2 \gg k_1$
$C_A = C_{AO}e^{-k_1 t}$
$C_R \approx 0$ (difficult to measure)
$\left(\dfrac{dC_R}{dt} = 0\right)$
$C_S = C_{AO}(1 - e^{-k_1 t})$

"One which is the slowest controls the overall rate of production of C_S"

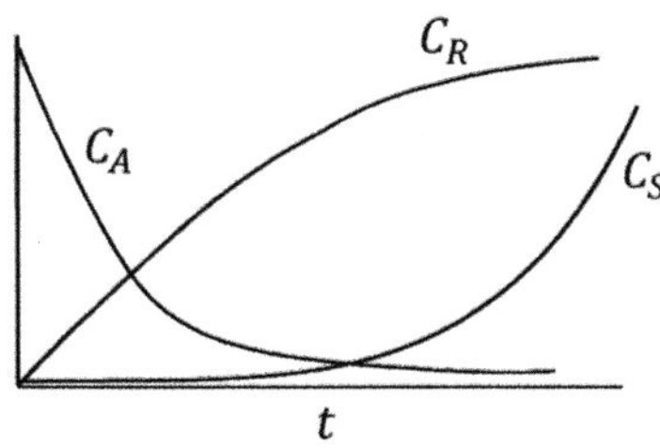 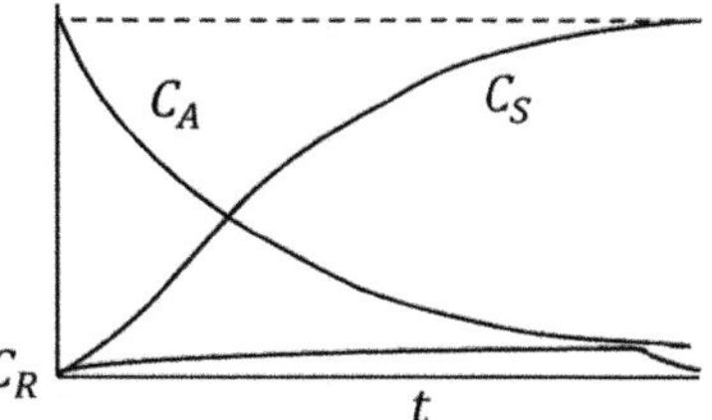

(g) 1st -order reversible reaction

$$A \underset{k_2}{\overset{k_1}{\rightleftharpoons}} R \qquad K_c = K = \text{equilibrium const.} = \frac{k_1}{k_2}$$

$$M = \frac{C_{RO}}{C_{AO}}; \quad C_A = C_{AO}(1 - X_A)$$

$$\frac{dC_R}{dt} = -\frac{dC_A}{dt} = C_{Ao}\frac{dX_A}{dt} = k_1 C_A - k_2 C_R$$

$$= k_1(C_{AO} - C_{AO}X_A) - k_2(C_{RO} + C_{AO}X_A)$$

$$= k_1 C_{AO}(1 - X_A) - k_2 C_{AO}(M + X_A)$$

At equilibrium,

$$\frac{dC_A}{dt} = \frac{dC_R}{dt} = 0 \quad (r_f = r_b)$$

$$K(T) = \frac{k_1}{k_2} = \frac{M + X_{Ae}}{1 - X_{Ae}} = \frac{C_{Re}}{C_{Ae}}$$

$K(T)$ is available from thermodynamic data.

Data:

t	C_A	C_R
0	50	0
1	40	$\vdots$
2	30	$\vdots$
$\vdots$	$\vdots$	$\vdots$
$\vdots$	10	30
∞	10	30
	C_{Ae}	C_{Re}

$$\frac{dX_A}{dt} = \frac{k_1(M+1)}{M+X_{Ae}}(X_{Ae} - X_A)$$

$$= \frac{C_{AO}X_{Ae}}{C_{Ae}(1-X_{Ae})}$$

$$-\ln\left(1 - \frac{X_A}{X_{Ae}}\right) = -\ln\frac{C_A - C_{Ae}}{C_{AO} - C_{Ae}}$$

$$= \left(\frac{(M+1)}{M+X_{Ae}}\right)k_1 t - k't$$

Get k_1 from here and k_2 from the ratio (k_1/k_2).

For irreversible reaction,

$$\left.\begin{array}{c} k_2 \to 0 \\ K \to \infty \\ X_{Ae} \to 1 \end{array}\right\}$$

In the limit $k_2 \to 0$, $K(T) \to \infty$, $X_{Ae} \to 1$, $C_{Ae} \to 0$.

or,

$$\frac{k_1(M+1)}{M+X_{Ae}} = \frac{k_1(M+1)}{M+1} = k_1$$

You have the same expression as before.

(h) Reaction of Shifting Order (Non-elementary Reaction)

$A \rightarrow R;$

$$-r_A = -\frac{dC_A}{dt} = \frac{k_1 C_A}{1 + k_2 C_A} = \frac{k_1}{\frac{1}{C_A} + k_2}$$

If

$$C_A \uparrow, \quad -r_A = \frac{k_1}{k_2} \quad \text{(Zeroth order)}$$

$$C_A \downarrow, \quad -r_A = k_1 C_A \quad \left(k_2 \ll \frac{1}{C_A}; C_A \ll \frac{1}{k_2} \right)$$

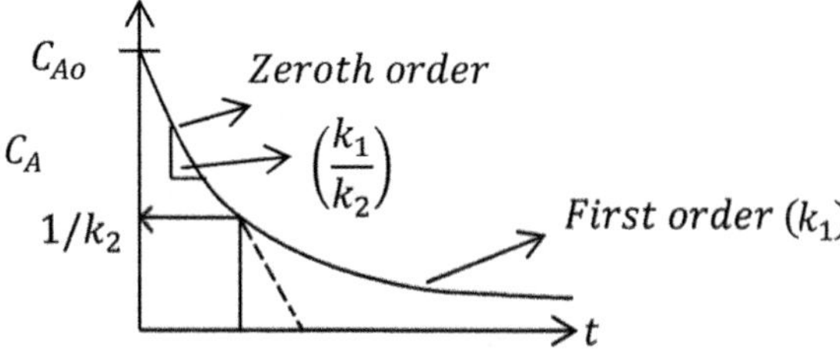

Enzyme or microbial reaction - Monod's kinetics shows shifting-order

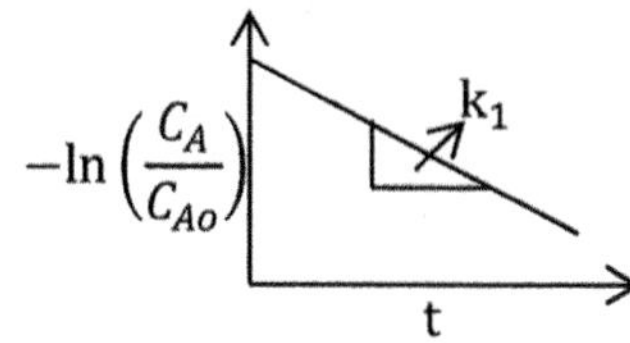

© The Author(s) 2025
N. Verma, *Chemical Reaction Engineering*,
https://doi.org/10.1007/978-3-031-88691-1_5

$$\ln \frac{C_{Ao}}{C_A} + k_2 \left(C_{Ao} - C_A \right) = k_1 t \quad \text{(full equation)}$$

$$\frac{\ln(C_{Ao}/C_A)}{C_{Ao} - C_A} = -k_2 + \frac{k_1 t}{C_{Ao} - C_A}$$

(get k_1 from slope and k_2 from intercept.)

(i) Reaction of nth order (empirical) and half life:

$$- r_A = - \frac{dC_A}{dt} = k C_A^n$$

$$C_A^{1-n} - C_{Ao}^{1-n} = (n-1)kt, \quad n \neq 1$$

Note:

(1) $n \gg 1$; $C_A \to 0$ if $t \to \infty$

(2) $n < 1$; $C_A = 0$ at $t \geq \frac{C_{Ao}^{1-n}}{(1-n)k}$ (0 or fraction)

(3) For

$$C_A^{1-n} - C_{Ao}^{1-n} = (n-1)kt$$

$C_A = \frac{C_{Ao}}{2}$ at $t = t_{1/2}$ (useful approach for studying radioactive elements)

$$t_{\frac{1}{2}} = \frac{(0.5)^{1-n} - 1}{k(n-1)} C_{Ao}^{1-n}$$

(get k from here if 'n' is known.)

Else,

$$\log\left(t_{\frac{1}{2}}\right) = \log\left(\frac{(0.5)^{1-n} - 1}{k(n-1)}\right) + (1-n)\log C_{A0},$$

$$n \neq 1 \begin{cases} \text{slope} \to n \\ \text{intercept} \to k \end{cases}$$

That is, use different C_{Ao}:

$$\left.\begin{array}{cc} C_{A01} & t_{\frac{1}{2}} \\ C_{A02} & t_{\frac{1}{2}} \\ C_{A03} & t_{\frac{1}{2}} \end{array}\right\} t_{\frac{1}{2}}$$

C_{Ao}	100	90	80	...	60	...	...
$t_{1/2}$	...	...	...	...	...	...	...

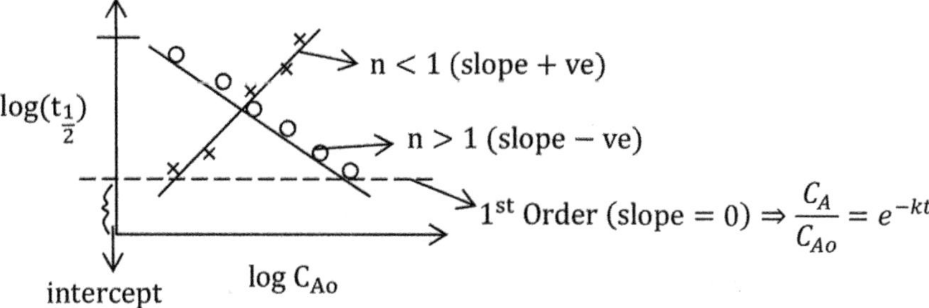

$$\frac{C_A}{C_{Ao}} = e^{-kt}$$

(j) $\alpha A + \beta B + \cdots \to$ Products. If A and B... are taken in the stoichiometric ratio $\alpha : \beta \ldots$ then their concentrations will remain in the same ratio during reaction.

$$-r_A = kC_A^\alpha C_B^\beta \ldots$$
$$= k'C_A^{\alpha+\beta+\ldots}$$

(i.e., $C_B : C_A = \beta : \alpha$)

$$-\frac{dC_A}{dt} = k'C_A^{\alpha+\beta+\ldots}$$

(then k' or k can be determined same as before.)

Varying volume batch reactor. (gas at constant P)

$$-r_A = -\frac{1}{V}\frac{dN_A}{dt} \neq -\frac{dC_A}{dt} \quad (\text{because } V = V(t))$$

$$V = V_o \left(1 + \epsilon_A X_A\right) \Rightarrow \text{linearly related } - \text{ assumption}$$

$$dV = V_o \epsilon_A dX_A$$

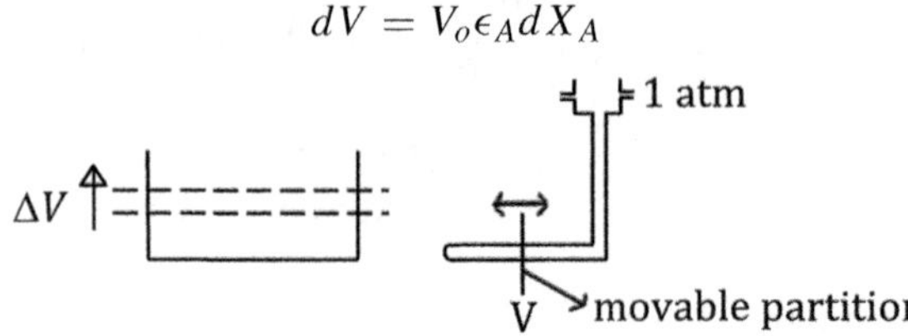

where,

$$\epsilon_A = \left(\frac{V_{X_A=1} - V_{X_A=0}}{V_{X_A=0}} \right) - \text{def}^n$$

1. $A \rightarrow 4R$, for example

$$\epsilon_A = \frac{4 - 1}{1}$$
$$= 3$$
$$V = V_o \left(1 + 3X_A\right)$$

2. $A \rightarrow 4R$ (50% inert)

$$\epsilon_A = \frac{(2 + 0.5) - 1}{1}$$
$$= 1.5$$
$$= \frac{(4 + 1) - 2}{2}$$
$$= 1.5$$

$$N_A = N_{Ao} \left(1 - X_A\right) \text{ still holds good; } V = V_o \left(1 + 1.5 X_A\right)$$

$$C_A = C_{Ao} \left(1 - X_A\right); \epsilon_A = 0$$

t	$C_A \rightarrow X_A \rightarrow V$
0	⋮
1	⋮
2	⋮
⋮	⋮

$$C_A = \frac{N_A}{V} = \frac{N_{Ao}\left(1 - X_A\right)}{V_o\left(1 + \epsilon_A X_A\right)} = C_{Ao}\frac{\left(1 - X_A\right)}{\left(1 + \epsilon_A X_A\right)}$$

$$-r_A = -\frac{1}{V}\frac{dN_A}{dt}\left(\neq \frac{dC_A}{dt}\right) = \frac{C_{Ao}}{(1+\epsilon_A X_A)}\frac{dX_A}{dt}$$

$$\left.\begin{array}{l} V = V_o\left(1+\epsilon_A X_A\right) \\ dV = V_o\epsilon_A dX_A \end{array}\right\}$$

Data:

t	$V \rightarrow$	X_A or C_A or p_A
$\vdots$	V_o	$\vdots$
$\vdots$	$\vdots$	$\vdots$
$\vdots$	$\vdots$	$\vdots$
$\vdots$	$\vdots$	$\vdots$

$(p_A = C_A RT \text{ for gas})$

$$\begin{aligned} -r_A &= \frac{C_{Ao}}{(1+\epsilon_A X_A)}\frac{dX_A}{dt} \\ &= \frac{C_{Ao}}{V\epsilon_A}\frac{dV}{dt} \\ &= \frac{C_{Ao}}{\epsilon_A}\frac{d(\ln V)}{dt} \end{aligned}$$

Fit the data to extract the rate constant.

Gaseous reaction (Constant Volume)

$$A \longrightarrow 2B \text{ (pressure will increase)}$$

$$-r_A = -\frac{dC_A}{dt} = -\frac{1}{RT}\frac{dp_A}{dt} = kC_A^n = k/(RT)^n p_A^n$$

t	P (atm)	p_A (atm)	
0	1		
$\vdots$	1.1		use stoichiometry
$\vdots$	$\vdots$		
30	2		

Differential Analysis: Work on the rate $(-r_A)$ itself. If the rate expression is difficult to integrate using mathematical methods or analytical techniques, consider finding derivatives (dC_A/dt) from the experimental measurements of concentration and calculate $(-r_A)$.

t	C_A	t	$-r_A$
0	C_{A0}	t_1	$-r_{A_1}$
t_1	C_{A1}	t_2	$-r_{A_2}$
t_2	C_{A2}	t_3	$-r_{A_3}$

Example: Reaction of Shifting order:

$$- r_A = -\frac{dC_A}{dt} = \frac{k_1 C_A}{1 + k_2 C_A}$$

$$\left(\frac{1}{-r_A}\right) = \frac{1}{k_1 C_A} + \left(\frac{k_2}{k_1}\right)$$

C_A	r_A
⋮	⋮
⋮	⋮

Linear Regression. (Least Square Fit for $z_n = f(x_n, y_n)$)
Equation:

$$1 + \alpha x_n + \beta y_n - y z_n = \epsilon_n$$

for example, $(-r_A) = f(T, C_A)$.

$\sum_{n=1}^{N} \epsilon_n^2$ should be minimized. Data:

x_1	y_1	z_1
⋮	⋮	⋮
x_n	y_n	z_n

$$\alpha \sum x_n^2 + \sum x_n + \beta \sum x_n y_n - y \sum x_n z_n = 0$$

$$\beta \sum y_n^2 + \sum y_n + \alpha \sum x_n y_n - y \sum y_n z_n = 0$$

$$- y \sum z_n^2 + \sum z_n + \alpha \sum x_n z_n - \beta \sum y_n z_n = 0$$

(1)	(2)	(3)	(4)	(5)	(6)	(7)	(8)	(9)
x_1	y_1	z_1	$x_1 y_1$	$x_1 z_1$	$y_1 z_1$	x_1^2	y_1^2	z_1^2
⋮	⋮	⋮	⋮	⋮	⋮	⋮	⋮	⋮
x_n	y_n	z_n	⋮	⋮	⋮	⋮	⋮	⋮
$\sum x_n$	$\sum y_n$	$\sum z_n$	$\sum x_n y_n$	$\sum x_n z_n$	$\sum y_n z_n$	$\sum x_1^2$	$\sum y_1^2$	$\sum z_1^2$

3 unknown and 3 equations to solve for α, β & γ.

If isothermal, only two variables (x_n, y_n):

$$\begin{cases} \alpha \sum x_n^2 + \sum x_n + \beta \sum x_n y_n = 0 \\[2mm] \beta \sum y_n^2 + \sum y_n + \alpha \sum x_n y_n = 0 \\[2mm] 1 + \alpha x_n + \beta y_n = 0 : \text{ Equation to be fitted} \end{cases}$$

Rate equation: for example,

$$\ln(C_A/C_{Ao}) = -kt \Rightarrow \ln C_A - \ln C_{Ao} + kt = 0$$

$$\frac{\ln(C_A)}{\ln(C_{Ao})} - 1 + \left(\frac{k}{\ln C_{Ao}}\right) t = 0$$

$$1 - \underbrace{\frac{\ln C_A}{\ln C_{Ao}}}_{x_n} - \left(\frac{k}{\ln C_{Ao}}\right) \underbrace{t}_{y_n} = 0$$

or,

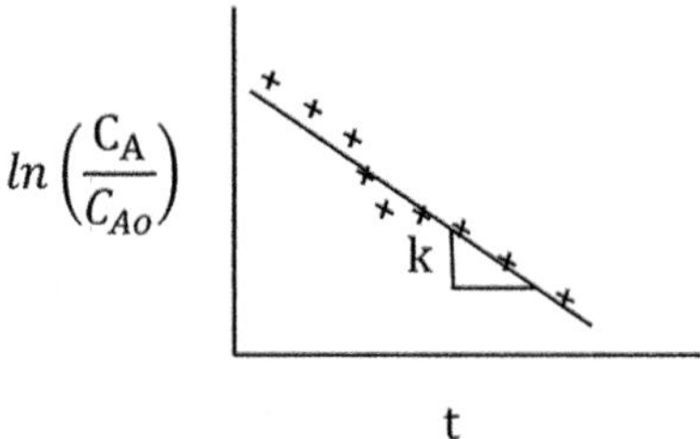

Example 1

Consider the following kinetic rate expression used to explain a liquid-phase enzymatic reaction:

$$- r_A = \frac{k_1 C_A C_{EO}}{C_A + k_2}; \quad \text{moles/s-lt}$$

where, C_{EO} = (initial) enzyme concentration = 0.01 milli-mol/lt,

$$k_1, k_2 \text{ are the model kinetic rate parameters}$$

$$C_A \equiv \text{glucose concentration (milli-mol/lt)}$$

The batch kinetic study records the following data under isothermal condition

t (h)	1	3	5	7	9	11
C_A	0.84	0.53	0.27	0.09	0.018	0.0025

Determine k_1 and k_2; C_{AO} = 1.0 milli-mol/lt.

Ans:

(1) It is clear that we are discussing a non-elementary reaction, as

$$(-r_A) \neq f(T)\phi(C_A)$$

(2) The rate expression can be analytically integrated;

$$-\frac{dC_A}{dt} = \frac{k_1 C_A C_{EO}}{k_2 + C_A} \implies -\frac{dC_A}{dt} = \frac{k_1' C_A}{1 + k_2' C_A} \left(\begin{array}{l} k_1' = \frac{k_1}{k_2} C_{EO} \\[4pt] k_2' = \frac{1}{k_2} \end{array} \right)$$

to obtain

$$\ln\left(\frac{C_{AO}}{C_A}\right) + k_2'\left(C_{AO} - C_A\right) = k_1' t$$

This equation can also be arranged to write as

$$\frac{\ln(C_{AO}/C_A)}{C_{AO} - C_A} = -k_2' + \frac{k_1' t}{C_{AO} - C_A}$$

(3) You can use any form to cast this equation in a linear equation: $y = mx + c$ and fit the data to determine slope (m) and intercept (c).

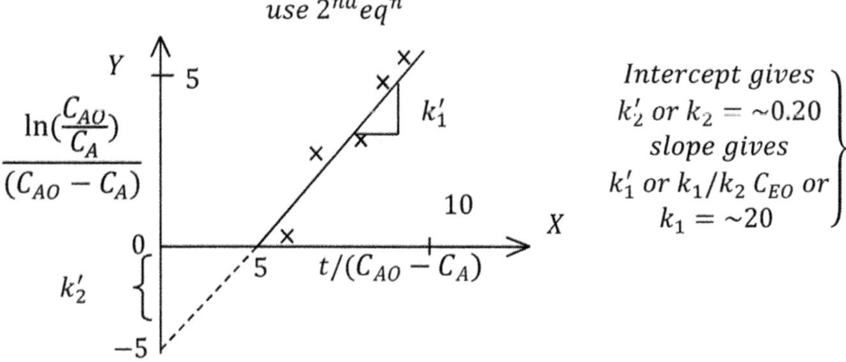

Note that the data have to be recalculated as;

t	✓	✓	✓	✓	✓	✓
$C_{AO} - C_A$	✓	✓	✓	✓	✓	✓
C_{AO}/C_A	✓	✓	✓	✓	✓	✓
$\left(\dfrac{C_{AO}/C_A}{C_{AO} - C_A}\right)$	✓	✓	✓	✓	✓	✓

(4) In general, you should analytically integrate the rate expression, if you can! Like here. Else, numerical technique is required, i.e., Trapezoidal rule of integration or Simpson's $1/3^{rd}$ rule etc. You need to write a code so that the numerically predicted values of C_A versus time using the different values of k_1 & k_2 fit the batch data. The important thing to note herein is that you can also use differential method to determine rate constants, etc., in which case one has to calculate (or use) the rate, i.e., $(-r_A)$;

$$-r_A = -\frac{dC_A}{dt}$$

from concentration data, using the central difference or forward/backward difference methods. It is a difficult choice; Numerical integration or Numerical differentiation? Given sufficient number of samples for analysis, any of two methods should be O.K. If the slopes or gradients $\left(\frac{dC_A}{dt}\right)$ are steep, you *may* prefer numerical integration. The best way to take such decision is to plot C_A versus **time** and inspect the slope qualitatively.

For the present case, if you apply differential method; the linearized form of equation takes the form $(y = mx + c)$

$$\left.\begin{array}{c} (-r_A) = -\dfrac{dC_A}{dt} = \dfrac{k_1' C_A}{1 + k_2' C_A} \Rightarrow \dfrac{1}{(-r_A)} = \dfrac{1}{k_1' C_A} + \dfrac{k_2'}{k_1'} \\[3mm] \text{or} \quad (-r_A) = \dfrac{k_1'}{k_2'} - \dfrac{1}{k_2'} \left[\dfrac{(-r_A)}{C_A}\right] \end{array}\right\}$$

Any of the two linear equations will work with the following data:

$$\left.\begin{array}{c} C_A \\[2mm] (-r_A) \\[2mm] 1/C_A \\[2mm] \dfrac{1}{(-r_A)} \\[2mm] -\dfrac{r_A}{C_A}, \; etc. \end{array}\right\} \quad \begin{array}{c} Fill-up\ yourself\ \&\ plot \\ to\ get\ slope\ \&\ intercept. \\ Check\ the\ final\ results\ for\ k_1\ \&\ k_2 \\ vis\ a\ vis\ those\ from\ numerical\ integration: \end{array}$$

Example 2 (From the Levenspiel-Book)

The following data are obtained at 0 °C in a constant-volume batch reactor using pure gaseous A:

Time (min)	0	2	4	6	8	10	12	14	∞
Partial pressure of A (mm)	760	600	475	390	320	275	240	215	150

The stoichiometry of the decomposition is $A \Rightarrow 2.5R$. Find a rate equation.

Ans: There are several take-away messages to consider before solving.

(1) We are discussing a constant-volume batch reactor for gas-phase reaction. There is clearly compressibility effort: $\rho = \rho\,(p)$ for gas, and a constant-pressure batch reactor can also be used to study gas-kinetics. For the liquid-phase, no such scenario exists as liquids are considered to be a constant density fluid.

(2) Check the stoichiometry, total pressure in the reactor will increase, unlike in the gaseous phase reaction like

$$H_2\,(g) + Br_2\,(g) \rightarrow 2HBr\ (g)\ (\Delta n = 0)$$

In the present case, total pressure can be monitored using a pressure-gauge, and the partial pressure of A or R can be calculated from stoichiometry. There is no need for the gas-sampling or chemical analysis to determine concentration, C_A or C_R, as $C_A = p_A RT$. However, if Δn or the total moles change is zero in a constant 'V' reactor at constant 'T', the gas-sampling must be done to determine $C_A = C_A\,(t)$, etc. On the other hand, for liquid-phase reaction, $C_A(t)$ is always determined experimentally.

(3) After a long time $(t \rightarrow \infty)$, there is an appreciable partial pressure of A (150 mm), implying we are discussing a reversible reaction $A \rightleftharpoons 2.5R$, with two constants $k_{forward}$ & $k_{reverse}$ (although no reaction is truly irreversible!). Coming to the solution, the stoichimetry is given and no rates are indicated here. Thus, a non-elementary reaction can assume any form!

Let us make this problem simple by assuming a $1st$-order reaction rate

$$-\frac{dC_A}{dt} = k_1\ C_A - k_2 C_R$$

(Use 'C_A' instead of 'p_A' as much as possible even if we discuss gas-phase kinetics)

$$C_R = C_{RO} + 2.5\,(C_{AO} - C_A) = 2.5\,(C_{AO} - C_A)$$

$$C_{Re} = 2.5\,(C_{AO} - C_{Ae});\quad \frac{k_1}{k_2} = \frac{C_{Re}}{C_{Ae}}\ \left(At\ equilbrium - \frac{dC_A}{dt} = 0\right)$$

Use these quantities to show

$$-\frac{dC_A}{dt} = \frac{k_1 C_{AO}\,(C_A - C_{Ae})}{(C_{AO} - C_{Ae})}\quad \text{(There is a cancellation of '2.5'!)}$$

Integrate

$$\ln\left(\frac{C_A - C_{Ae}}{C_{AO} - C_{Ae}}\right) = \left(\frac{C_{AO}}{C_{AO} - C_{Ae}}\right) k_1 t$$

or

$$\ln\left(\frac{p_A - p_{Ae}}{p_{AO} - p_{Ae}}\right) = \left(\frac{p_{AO}}{p_{AO} - p_{Ae}}\right) k_1 t;\quad p_{Ae} = 150\ mm,\ p_{AO} = 760\ mm$$

Use the data as;

t (min)	0	2	...	14	∞
$\ln\left(\dfrac{p_A - p_{Ae}}{p_{AO} - p_{Ae}}\right)$	0	-0.304	...	-2.239	$-\infty$

Plot $y = mx + c \Rightarrow$ the slope will give $k_1 = 0.1295\,\text{min}^{-1}$

$$k_2 = k_1\left(\frac{C_{Ae}}{C_{Re}}\right) = 0.1295 \times \frac{150}{2.5(760 - 150)} = 0.0127\,\text{min}^{-1}$$

$$-r_A = 0.1295 C_A - 0.0127 C_R$$

Another take-away: For any reaction other than the $1st$-order, be careful while working with pressure-unit for the rate constant. From the kinetics point of view,

$$-\frac{dC_A}{dt} = kf(C_A)$$

is intrinsically correct. On the other hand,

$$-\frac{dp_A}{dt} = kf(p_A)$$

assumes a transformed form in pressure units. Also, it is our convenience of measurements of p or p_A for gas that we prefer to use 'p_A' for calculation by using

$$C_A = \frac{p_A}{RT}$$

Develop a habit of working on C_A, instead of p_A.

Example 3

A first-order homogeneous gaseous decomposition $A \rightarrow 2.5R$ is carried out in an isothermal batch reactor at 2 atm with 20% inert present, and the volume increases by 60% in 20 min. In a constant-volume reactor, find the time required for the pressure to reach 8 atm if the initial pressure is 5 atm, 2 atm of which consists of inerts?

Ans: This problem involves both constant pressure- (varying volume) and constant volume- (varying pressure) operation of a batch reactor. Here, we are referring to a gas-phase reaction ($\rho = \rho(P)$) and an especially designed varying volume reactor (possibly, a flexible and not a rigid-wall reactor).

Part 1:   (Varying volume-reactor; constant pressure)

$$\varepsilon_A = \frac{(2.5 \times 0.8 + 0.2) - (0.8 + 0.2)}{(0.8 + 0.2)} = 1.2$$

$$\underset{A \quad \text{inert}}{\uparrow \quad \uparrow}$$

(here, basis is 1 mole of the reaction mixture)

$$- r_A = \frac{-1}{V} \frac{dN_A}{dt} \left(\neq \frac{dC_A}{dt} \text{ as } V = V(t) \right) = k_1 C_A$$

$$V = V_o (1 + \varepsilon_A X_A) \implies X_A = \frac{1.6 - 1}{1.2} = 0.5 \text{ in 20 min}$$

$$N_A = N_{Ao} (1 - X_A) \implies C_A = \frac{C_{Ao} (1 - X_A)}{(1 + \varepsilon_A X_A)}$$

Substituting,

$$+ \frac{N_{Ao}}{V_o(1 + \varepsilon_A X_A)} \frac{dX_A}{dt} = k_1 \frac{C_{Ao} (1 - X_A)}{(1 + \varepsilon_A X_A)}$$

or

$$\frac{dX_A}{dt} = k_1(1 - X_A) \implies -\ln(1 - X_A) = k_1 t$$

(*Note:* the expression is the same as that for the batch reactor at constant pressure! Had this been any reaction-order other that the $1st$-order, the rate expression would have been different, showing 'ε_A' effect).

$$k_1 = - \frac{\ln(1 - 0.5)}{20} = 0.0346 \text{ min}^{-1}$$

Part 2:   (Constant volume-reactor; pressure varies)

$$- r_A = - \frac{1}{V} \frac{dN_A}{dt} = - \frac{dC_A}{dt} = k_1 C_A$$

$$- \frac{dp_A}{dt} = k_1 p_A \quad (p_A = C_A RT)$$

$$\frac{dX_A}{dt} = k_1 (1 - X_A) \implies -\ln(1 - X_A) = k_1 t$$

	Initial	**Final**
A:	3	$(3(1 - X_A))$
Inert:	2	2
R:	0	$2.5 \times (3X_A)$
		$\uparrow$ converted $3(1 - X_A) + 2 + 7.5 X_A = 8$ atm $X_A = 0.666$

Substituting, $-\ln(1 - 0.666) = 0.0346t \implies t = 31.69$ min.

Some take-away: For gas-system, it is strongly recommended to work on 'C_A' rather that 'p_A'. For the $1st$-order reactions, the rate constant & activation energy calculations remain unchanged. However, for any other order, 'k' using C_A may not be the **same as** 'k_p' using 'p_A'. Similarly, there may be fallacy in calculating 'E' using 'k_p' and 'p_A'. See the solved example in the text book by Levensipiel.

Example 4

A batch reactor is used to study the behavior of the following irreversible liquid-phase reaction in series $A \to R \to 2S$. Both reactions are first-order. The following time (min)-concentration (mol/m^3) data are measured for A and S.

Time (min)	0	1	4	7	10	15	20
A (mol/m^3)	100	75	45	20	10	3	1
S (mol/m^3)	0.00	4.3	45	92	130	170	187

Determine the rate constants. The initial concentration of R is 0.

Ans: Inspect carefully. We have $A \xrightarrow{k_1} R \xrightarrow{k_2} 2S$ and not $A \xrightarrow{k_1} R \xrightarrow{k_2} S$. Be careful in copying the rate- expression and solution from the book or elsewhere for the latter type of reaction.

First calculate $C_R(t)$. Stoichiometry:

$$C_A + C_R + C_S/2 = \text{constant} = C_{Ao} = 100 \frac{\text{mol}}{\text{m}^3}$$

will work and not $C_A + C_R + C_s = C_{Ao}$. It may be obvious to you, but a fresh calculation is recommended:

$$-\frac{dC_A}{dt} = k_1 C_A; \quad \frac{dC_R}{dt} = k_1 C_A - k_2 C_R; \quad \frac{dC_s}{dt} = 2k_2 C_R$$

Arrange:

$$\frac{d}{dt}(C_A + C_R + C_S/2) = 0 \Rightarrow C_A + C_R + C_S/2 = \text{constant}$$

$$\neq f(t) = C_{Ao} + C_{Ro} + C_{So}/2 = C_{Ao} = 100\,\frac{mol}{m^3}$$

(same as that from stoichiometry)

$$C_R = 100 - C_A - C_S/2$$

t	0	1	4	7	10	15	20
C_A	100	75	45	20	10	3	1
C_s	0.0	4.30	44.97	92.42	130.33	169.02	187.07
C_R	0	22.85	32.51	33.79	24.83	12.48	5.46
$\ln(C_{Ao}/C_A)$	0	0.287	0.798	1.609	2.302	3.506	4.605

Plot:

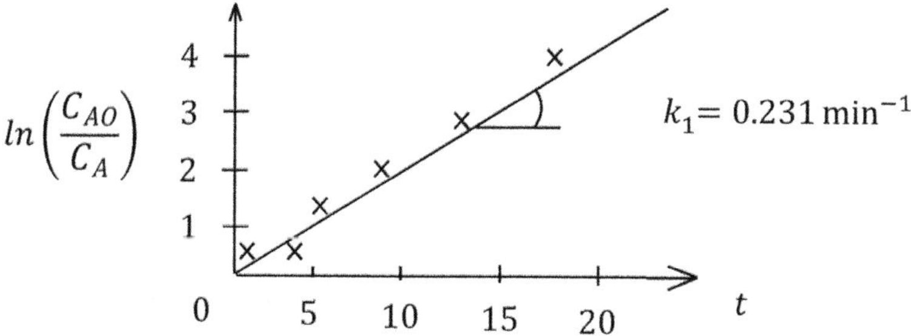

How do you determine k_2? You can substitute the solution for C_A in the rate expression for C_R and integrate to obtain

$$C_R = C_{Ao}k_1\left(\frac{e_1^{-k_1 t}}{k_2 - k_1} + \frac{e_1^{-k_2 t}}{k_1 - k_2}\right)\ (k_1 \neq k_2)$$

and then linearize? Then, from slope or intercept get k_2. Such approach will be full of error and is not required. In fact, you should first inspect the behavior of 'C_R'.

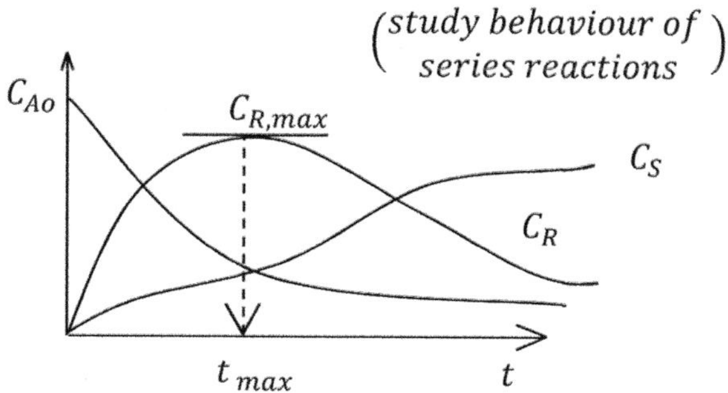

Recall, C_R is max at time

$$t_{max} = \frac{1}{\text{log-mean avg of } (k_1 \text{ and } k_2)}$$

$$= \frac{\ln (k_1/k_2)}{k_1 - k_2}$$

$$= \frac{\ln (0.2317/k_2)}{0.2317 - k_2} = 7 \text{ min} \quad \text{(from data)}$$

By iterations or Newton-Raphson Method (!)

$$k_2 = 0.133 \text{ min}^{-1}$$

Alternatively, $C_{R,max}$ at

$$\frac{dC_R}{dt} = 0 = k_1 C_A - k_2 C_R$$

$$\Rightarrow k_1 C_{Ao} \exp (-k_1 t) - k_2 C_R = 0$$

from which we can also get

$$k_2 = \frac{0.2317 C_{Ao} \exp (-0.2317 \times 7)}{34} = 0.133 \text{ min}^{-1}.$$

Reactor Design—Introduction (Mixed Flow Reactor)

Reactor Design

1. Batch–Tons, litre, purity (conversion), Yield (specification)
2. Flow reactors–TPD, m^3/hr, lpm *etc.*, purity (conversion), Production rate (specification)

Given feed rate, kinetics, mechanism, production rate, purity of product, no back **What is the reactor size?** (m^3, litre)

Steady state (start-up not considered) operation with respect to all variables (flow, concentration, temperature)

Flow Reactors

There are two model (ideal) reactors:

(1) CSTR (Continuous stirred tank reaction); MFR (Mixed flow reactor)
(2) PFR (Plug flow reactor).

© The Author(s) 2025

N. Verma, *Chemical Reaction Engineering*,

https://doi.org/10.1007/978-3-031-88691-1_6

CSTR

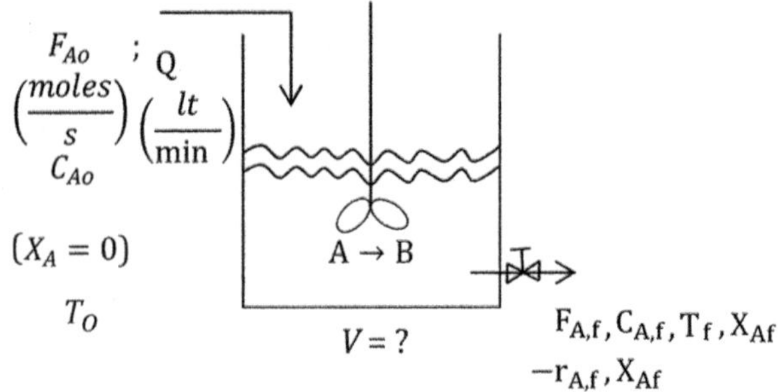

Assumption: $C_A, X_A \neq f(space)$ = Conditions at the outlet. The assumption is attributed to an uniform and instant mixing of reaction mixture inside the reactor. There is only one concentration level in CSTR; no progression of reaction in space (from inlet to outlet); 100% back-mixing.

PFR

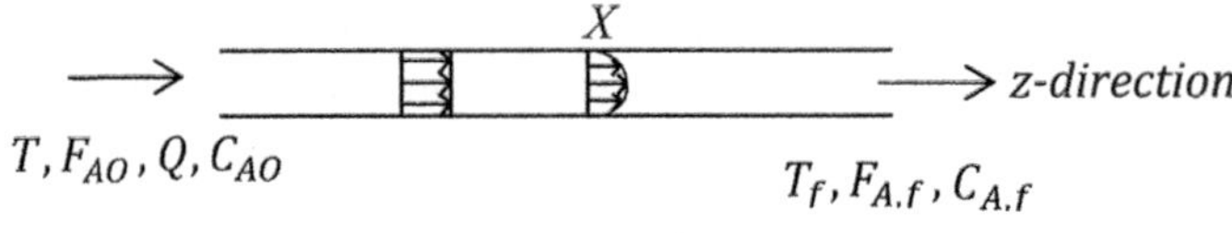

No back–mixing; zero mixing in flow (axial) direction but infinite mixing in r-direction. $C_A = C_A(z)$; $C_A \neq f(r)$. There is a progression of reaction in 'z'-direction.

Note:

$$Re = \frac{\forall d \rho}{\mu} = \frac{Q}{\frac{\pi}{4}d} \cdot \frac{\rho}{\mu} > 2100 \text{ in a tube}$$

(100% segregated flow) $\tau = V/v_0$ ($\varepsilon = 0$); τ = space-time/residense time. v_0 in the '0' subscript signifies the gas flowrate at STP condition.

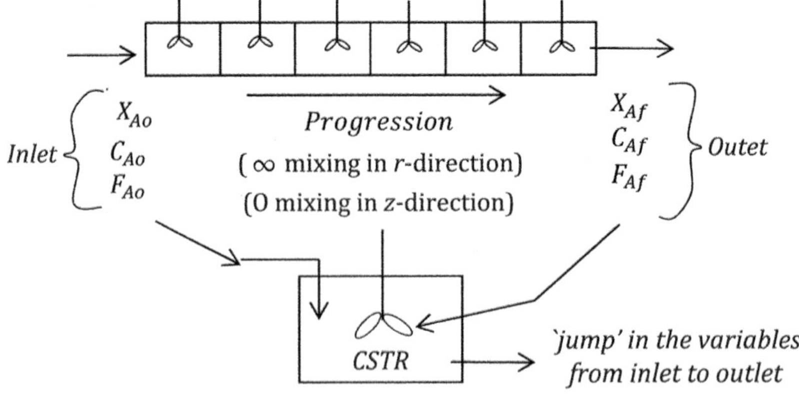

"PFR & CSTR are two extreme ends of the flow reactors," as far as mixing is concerned.

Plug flow

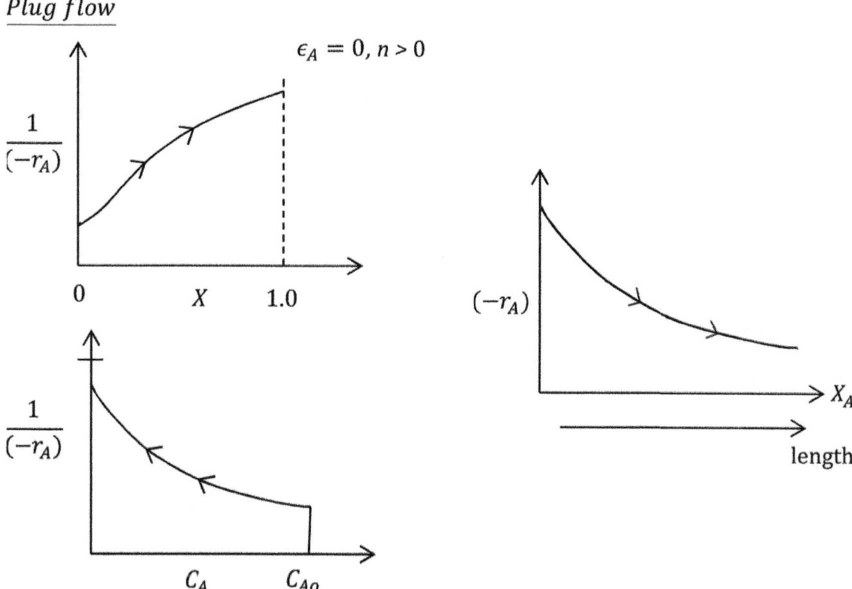

In batch (recall)

$$X_A(t) = \frac{N_{Ao} - N_A(t)}{N_{Ao}}$$

In flow

$$X_A = \frac{F_{Ao} - F_A}{F_{Ao}} \quad \text{-definition:}$$

$$F_A = F_{Ao}(1 - X_A)$$

$$dF_A = -F_{Ao}dX_A$$

(X_A must be defined with respect to the unreacted feed).

$$F_{Ao} = v_o C_{Ao} \quad , \quad F_{A,f} = v_f C_{Af}$$

$$moles/min \quad \downarrow$$

$$Lt/min \qquad mol/Lt$$

CSTR

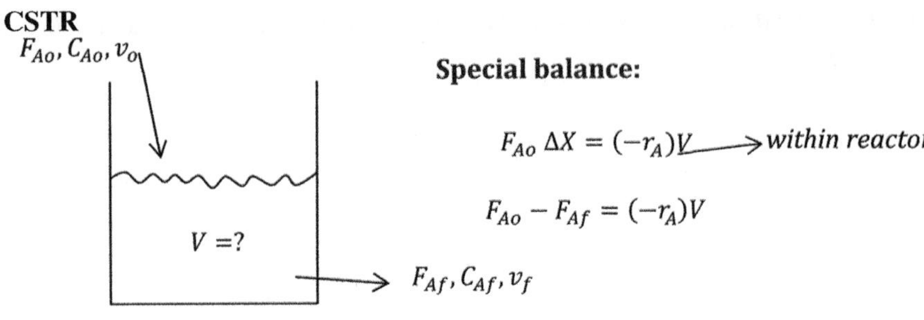

$$F_{Ao}X_f = (-r_A)_f V$$

$$v_o C_{Ao} X_f = (-r_A)_f V$$

Performance equation:

$$\frac{V}{v_o} = \tau = \frac{C_{Ao}X_{Af}}{(-r_A)_f} \tag{1}$$

τ = space time = residence time ($\epsilon = 0$)

$$= \frac{C_{Ao} - C_{Af}}{(-r_A)_f}, \text{ only if } \epsilon_A = 0$$

Similarly,

$$(-r_A)_f = kC_{Af} = kC_{Ao}\left(1 - X_{Af}\right), \quad \epsilon_A = 0, \quad \text{1st-order}$$

$$= \frac{kC_{Ao}\left(1 - X_{Af}\right)}{\left(1 + \epsilon_A X_{Af}\right)}, \quad \epsilon_A \neq 0, \quad \text{1st-order}$$

$$\left.\tau = \frac{V}{v_o} = \frac{C_{Ao}X_{Af}}{(-r_A)_f} = \frac{C_{Ao}X_{Af}}{kC_{Ao}\left(1 - X_{Af}\right)}\right\} \epsilon_A = 0, \quad \text{1st-order}$$

$$\left.\begin{array}{l} X_{Af} = \left(\dfrac{C_{Ao} - C_A}{C_{Ao} + C_A\epsilon_A}\right) \\[2mm] \tau = \dfrac{V}{v_o} = \dfrac{C_{Ao}X_{Af}\left(1 + \epsilon_A X_{Af}\right)}{kC_{Ao}\left(1 - X_{Af}\right)} \\[2mm] v_f = v_o\left(1 + \epsilon_A X_{Af}\right) \end{array}\right\} \epsilon_A \neq 0, \quad \text{1st-order}$$

Summary:

$$\left(\begin{array}{l} F_{Af} = F_{Ao}\left(1 - X_{Af}\right) \\[3mm] C_{Af} = \dfrac{C_{Ao}\left(1 - X_{Af}\right)}{\left(1 + \epsilon_A X_{Af}\right)} \\[3mm] v_f = v_o\left(1 + \epsilon_A X_{Af}\right) \end{array}\right)$$

Input variable: $F_{Ao}, C_{Ao}, v_o, X_{Af}$–Calculate τ or V from Eq. (1).

Graphically

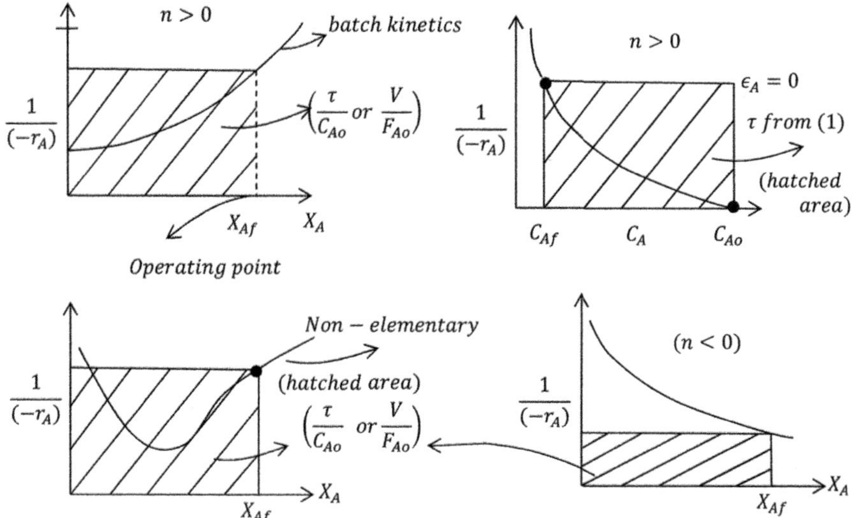

In an ideal CSTR-system, there are only two concentration levels; one at the inlet (feed) and the other at the exit, which is the same as within the reactor.

Plug Flow Reactor

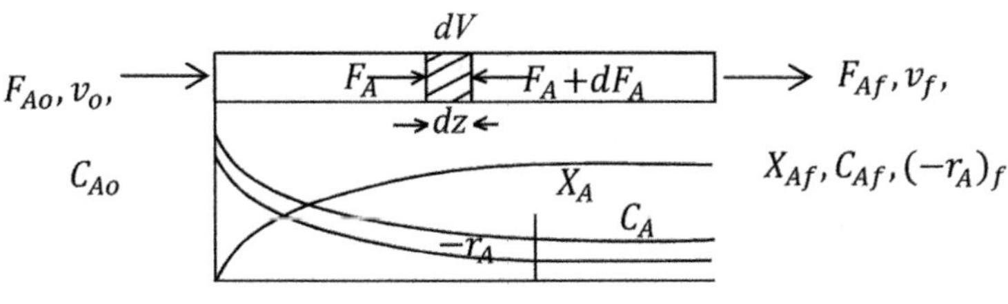

$$X_A = \frac{F_{Ao} - F_A}{F_{Ao}} - def^n$$

Species balance:

$$F_{Ao}dX_A = (-r_A)\,dV \text{ (moles/s)}$$

$$\left.\begin{aligned}
F_{A1} &= F_{Ao}\left(1 - X_{A1}\right) \\
F_{A2} &= F_{Ao}\left(1 - X_{A2}\right) \\
&\ \ \vdots \\
F_{Af} &= F_{Ao}\left(1 - X_{Af}\right)
\end{aligned}\right\} \text{ along the length}$$

Performance equation:

$$\frac{V}{F_{Ao}} = \int_0^{X_{Af}} \frac{dX_A}{(-r_A)}, \quad \epsilon \neq 0$$

Compare with CSTR:

$$\tau = \frac{C_{Ao}X_{Af}}{(-r_A)_f}$$

N. Verma, *Chemical Reaction Engineering*,
https://doi.org/10.1007/978-3-031-88691-1_7

or

$$\tau = C_{Ao} \int_0^{X_{Af}} \frac{dX_A}{(-r_A)}, \ \epsilon_A \neq 0$$

$$= -\int_{C_{Ao}}^{C_{Af}} \frac{dC_A}{(-r_A)}, \ \text{if } \epsilon_A = 0$$

$$\tau = \frac{C_{Ao} - C_{Af}}{(\overline{-r_A})}, \ \epsilon_A = 0; \ (\overline{-r_A}) = \text{avg. rate in PFR}$$

$$\left. \begin{array}{l} C_A = C_{Ao}(1 - X_A) \\ dC_A = -C_{Ao}dX_A \end{array} \right\} \varepsilon_A = 0$$

Note:

$$C_A = \frac{C_{Ao}(1 - X_A)}{(1 + \epsilon_A X_A)} \ \text{if } \varepsilon_A \neq 0$$

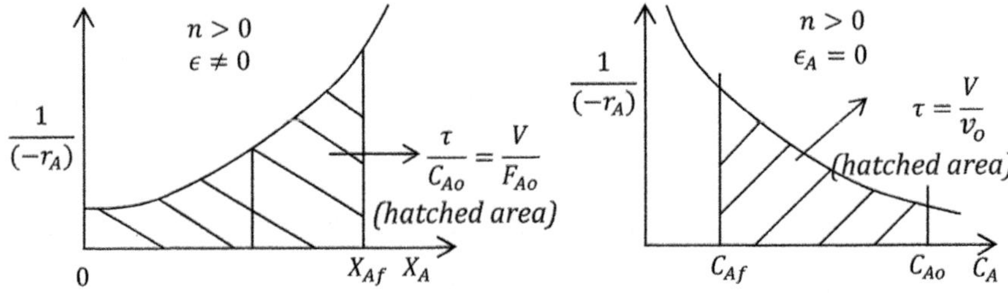

Size Comparison (CSTR Versus PFR)

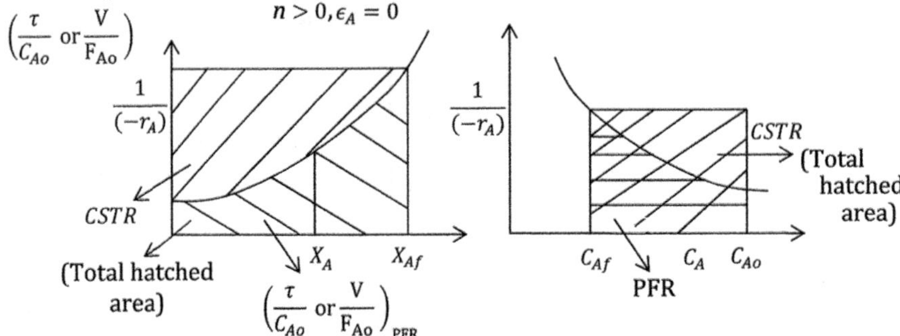

Conclusions:

(a) $V_{PFR} < V_{CSTR}$ for the same conversion and fluid flow rate, $n > 0$
(b) $n < 0$, $V_{PFR} > V_{CSTR}$
(c) $n = 0$, $V_{PFR} = V_{CSTR}$

Mechanistically, PFR with zero mixing in flow-direction sustains highest (maximum) level of concentration, therefore, maximum rate ($n > 0$) or the smallest reactor volume. Mixing lowers the concentration. Therefore, CSTR with infinite back-mixing has the lowest concentration or rate, and therefore, larger size or volume requirement for the same concentration.

PFR Versus Batch: $A \rightarrow$ **Products**

Batch data:

t	C_A
0	100
$\vdots$	$\vdots$
10 min	5

$$-\frac{dN_A}{dt} = (-r_A)V \rightarrow m^3$$
$$N_A = N_{Ao}(1 - X_A)$$
$$\frac{dX_A}{dt} = (-r_A)V$$

$$t = \frac{N_{Ao}}{V_o} \int_0^{X_A} \frac{dX_A}{(-r_A)(1 + \epsilon_A X_A)} \quad \text{(batch time)}$$

$$t = \frac{N_{Ao}}{V_o} \int_0^{X_A} \frac{dX_A}{(-r_A)}; \quad V \neq f(t) \text{ or } \varepsilon_A = 0$$

$$= C_{Ao} \int_0^{X_A} \frac{dX_A}{(-r_A)(1 + \epsilon_A X_A)}, \quad \epsilon_A \neq 0$$

$$= C_{Ao} \int_0^{X_A} \frac{dX_A}{(-r_A)}; \quad \varepsilon_A = 0$$

Plug flow reactor:

$$F_{Ao}dX_A = (-r_A)\,dV$$

$$\tau = C_{Ao} \int_0^{X_A} \frac{dX_A}{(-r_A)}; \quad \epsilon_A \neq 0 \text{ (general case)}; \tau = \frac{V}{v_0}$$

Compare with batch for $\epsilon_A = 0$:

$$t = C_{Ao} \int_0^{X_A} \frac{dX_A}{(-r_A)}; \quad \epsilon_A = 0$$

Therefore, batch time/holding time = space time (residence time) for the same conversion ($\varepsilon_A = 0$).

$\tau = t$

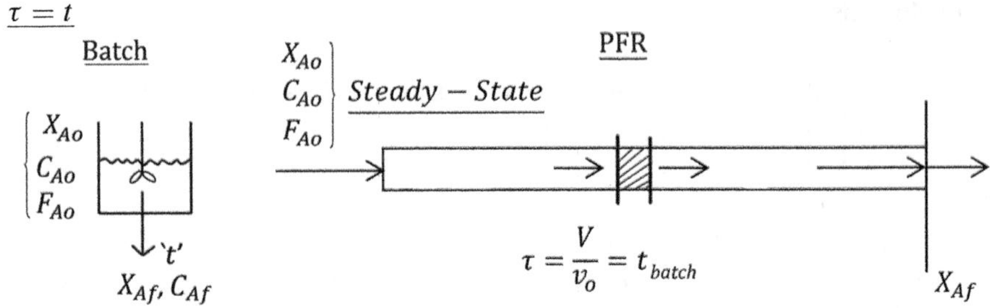

(1) If I require 5 min of batch time to process a feed to 'X' conversion, I need 5 min of holding time (or space time) in a PFR. Just as t in batch is independent of 'V', τ in plug flow is independent of F_{Ao} or v_o. It is 'V' which depends on τ for a given F_{Ao} or v_o.

(2) For $\varepsilon_A = 0$, $\tau_{PFR} = t_{batch}$ because flow is completely segregated with zero mixing in flow-direction. Therefore, a differential volume (ΔV) in PFR moves without mixing with its neighboring fluid element in flow-direction for the time τ_{PFR} as if the isolated volume spends t_{batch} time in batch reactor.

(3) Batch data can be interchanged with plug flow data ($\varepsilon_A = 0$).

(4) $t_{total} \neq t_{CSTR}$; flow is completely non-segregated; well mixed in all directions.

Space time versus mean residence time (τ versus $\bar{t}$)

τ = time needed to process one reactor volume of feed = $\dfrac{V}{v_o} = \dfrac{V C_{Ao}}{F_{Ao}}$

For example, 8 Lt of V & 2 lt/h $\Rightarrow$ 4 h to process one 'reactor-volume'.

$\dfrac{1}{\tau}$ = Space velocity = 0.25 h^{-1} = of reactor-volume processed in unit time.

(In one hour 0.25 reactor volume of feed will be processed)

$\bar{t}$ = mean residence time of flowing material

$$dt = \frac{dV}{v} = \frac{dV}{v_o(1 + \varepsilon_A X_A)}$$

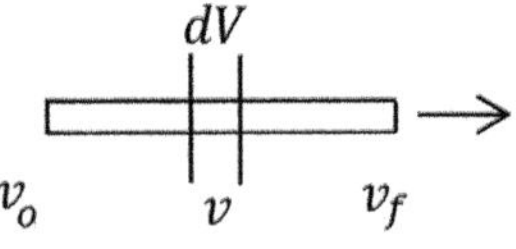

It is clear that $t \neq \tau$ if $\varepsilon_A \neq 0$ because there is an expansion or contraction of volume during the flow.

Mathematically,

$$\tau = C_{Ao} \int_0^{X_A} \frac{dX_A}{(-r_A)} \quad (PFR; \; \varepsilon \neq 0)$$

$$\bar{t} = C_{Ao} \int_0^{X_A} \frac{dX_A}{(-r_A)(1 + \epsilon_A X_A)}$$

Therefore,

$$\text{If} \epsilon_A > 0, \quad (A \to 2B)(\tau > \bar{t})$$

$$\text{If} \epsilon_A = 0, \quad \bar{t} = C_{Ao} \frac{V}{F_{Ao}} = \tau.$$

Single Reactor (Recall)

$$-r_A = -\frac{1}{V}\frac{dN_A}{dt} = kC_A^n$$

1. $F_{Ao}X_{Af} = (-r_A)_f\, V - \text{CSTR /MFR}$

$$
\begin{aligned}
\tau_m &= \left(\frac{V}{v_o}\right) \\
&= \left(\frac{VC_{Ao}}{F_{Ao}}\right) \\
&= \frac{C_{Ao}X_{Af}}{(-r_A)_f} \\
&= \frac{X_{Af}\left(1+\epsilon_A X_{Af}\right)^n}{kC_{Ao}^{n-1}\left(1-X_{Af}\right)^n}
\end{aligned}
$$

where, τ_m is design parameter. "Fix $\tau_m \rightleftharpoons$ Fix X_A" for a given C_{Ao}.

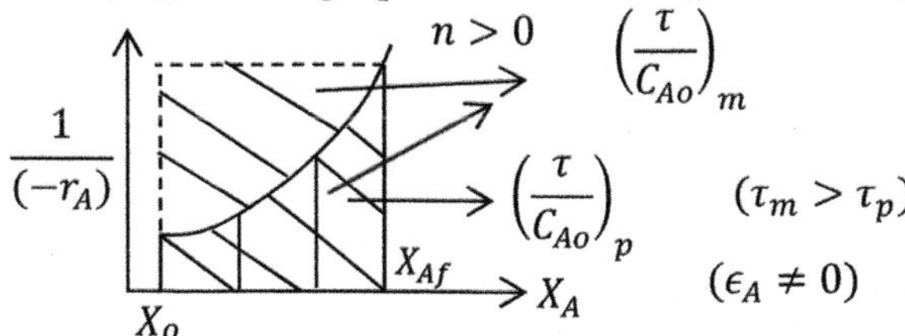

© The Author(s) 2025

N. Verma, *Chemical Reaction Engineering*,

https://doi.org/10.1007/978-3-031-88691-1_8

2. $F_{Ao} dX_A = (-r_A) dV$ (PFR)

$$\tau_p = \left(\frac{VC_{Ao}}{F_{Ao}}\right)_p = \frac{1}{kC_{Ao}^{n-1}} \int_0^{X_A} \frac{(1+\epsilon_A X_A)^n}{(1-X_A)^n} dX_A$$

$$\frac{\tau_m}{\tau_p} = \frac{V_m}{V_p} = \text{known } \left(f\left(X_f\right)\right)$$

$\underline{\epsilon_A = 0}$

$$n = 1: \quad \frac{(\tau)_m}{(\tau)_p} = \frac{\left(\frac{X_A}{1-X_A}\right)_n}{-\ln(1-X_A)_p}$$

$$\tau_m = \frac{C_{Ao} X_{Af}}{kC_{Ao}\left(1-X_{Af}\right)}, \quad \tau_p = -k\ln(1-X_A)$$

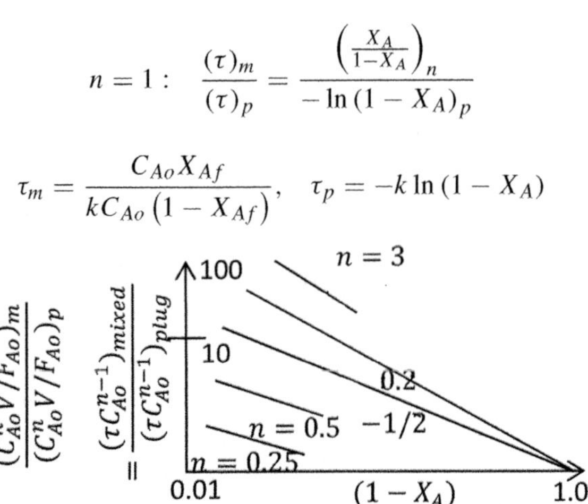

(1) The graphical results are available for size comparison of MFR & PFR for identical feeds (C_{Ao}, F_{Ao}) in terms of $\left(\frac{\tau_m}{\tau_p}\right)$, else

$$\frac{\left(c_{Ao}^n \frac{v}{F_{Ao}}\right)_m}{\left(C_{Ao}^n \frac{v}{F_{Ao}}\right)_p} \text{ on } y\text{-axis}$$

with/without variation of ϵ_A in case of $(-r_A = kC_A^n)$.

(2) For any particular duty & all +ve order reaction reaction, $\left(\tau_m > \tau_p\right)$ and $\left(V_m > V_p\right)$

(3) The ratio τ_m/τ_p increases for higher conversion (Y is log-axis).

(4) The ratio also increases with the order of reaction.

(5) ϵ_A (density effect) also causes the ratio to increase.

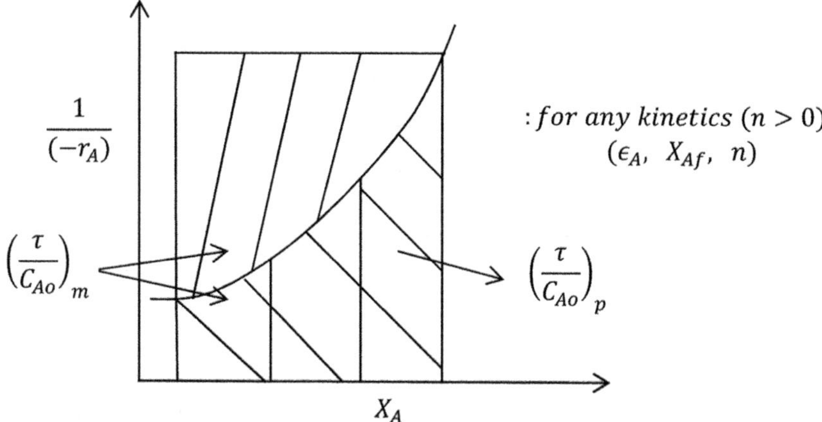

"In general for all reactions kinetics ($n > 0$), one should avoid micro-mixing, back-mixing, recycle, expansion, for minimum reactor size for the given conversion (X_{Af})".

Plug-Flow Reactors in Series

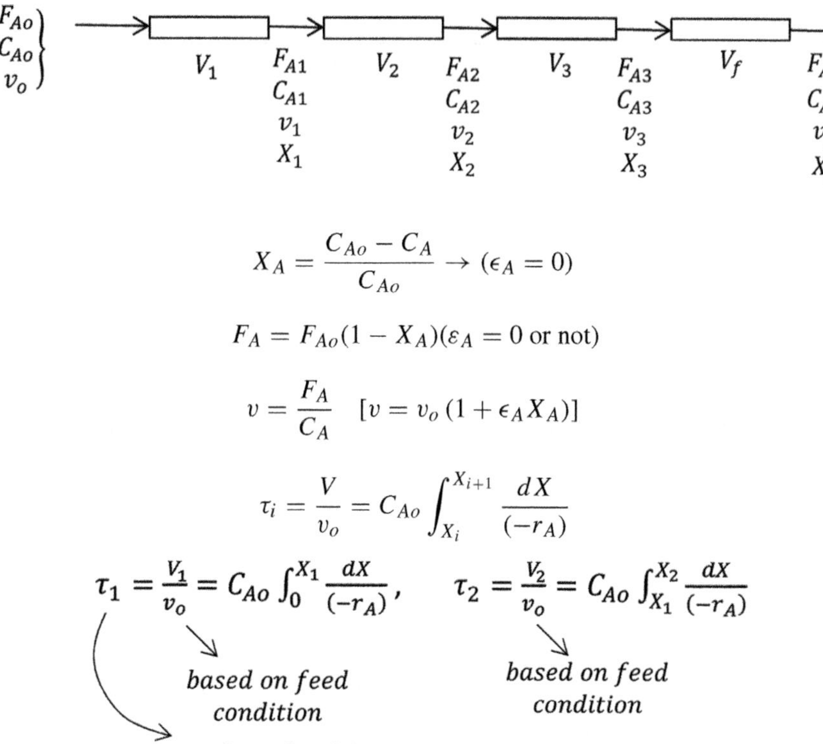

$$X_A = \frac{C_{Ao} - C_A}{C_{Ao}} \rightarrow (\epsilon_A = 0)$$

$$F_A = F_{Ao}(1 - X_A)(\varepsilon_A = 0 \text{ or not})$$

$$v = \frac{F_A}{C_A} \quad [v = v_o(1 + \epsilon_A X_A)]$$

$$\tau_i = \frac{V}{v_o} = C_{Ao} \int_{X_i}^{X_{i+1}} \frac{dX}{(-r_A)}$$

$$\tau_1 = \frac{V_1}{v_o} = C_{Ao} \int_0^{X_1} \frac{dX}{(-r_A)}, \qquad \tau_2 = \frac{V_2}{v_o} = C_{Ao} \int_{X_1}^{X_2} \frac{dX}{(-r_A)}$$

based on feed condition

based on feed condition

$$F_{Ao}dX = (-r_A)dV$$

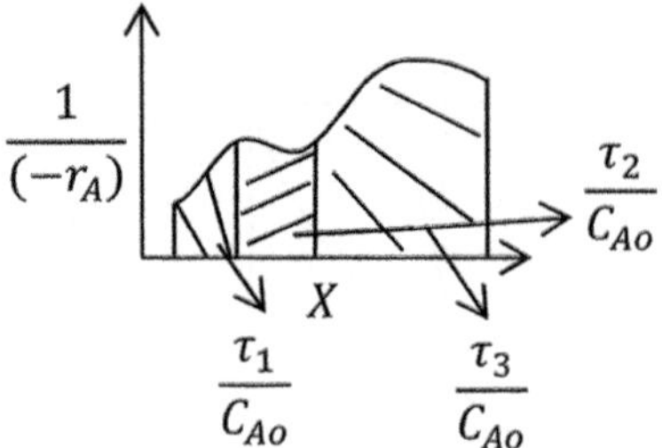

And also,

$$\tau = \frac{V}{v_o} = C_{Ao} \int_0^{X_{Af}} \frac{dX}{(-r_A)}$$

Thus, N plug flow reactors in series will give the same conversion as a single plug reactor of Volume V, with $V = \sum V_i$ or $\tau = \sum \tau_i$.

Parallel:

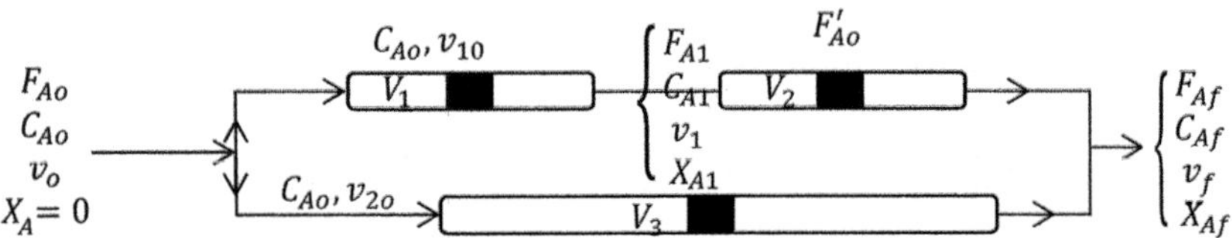

$$\left.\begin{array}{l} v_0 = v_{10} + v_{20} \\ F_{Ao} = F_{A1o} + F_{A2o} \end{array}\right\}$$

Notes:

(1) Splitting of the stream does not alter the concentration (conversion) of the split streams.
(2) Mixing of streams having different concentrations is not permissible (bad design). The concentrations of the streams post-reactors must be the same before mixing or rejoining.

Bottom stream:

$$F_{Ao}dX = (-r_A)\,dV$$

$$\tau_3 = \frac{V_3}{v_{20}} = C_{Ao} \int_0^{X_{Af}} \frac{dX}{(-r_A)_f}$$

$$C_{Af} = \frac{C_{Ao}\left(1 - X_{Af}\right)}{\left(1 + \epsilon_A X_{Af}\right)}$$

$$\frac{\text{moles}}{lt}$$

Top stream:

$$\tau_{top} = C_{Ao} \int_0^{X_{Af}} \frac{dX}{(-r_A)}$$

$$\tau_1 + \tau_2 = C_{Ao} \int_0^{X_{A1}} \frac{dX}{(-r_A)} + C_{Ao} \int_{X_{A1}}^{X_{Af}} \frac{dX}{(-r_A)}$$

$$\tau_1 + \tau_2 = \tau_{top} = \frac{V_1 + V_2}{v_{10}}$$

$$= \tau_3$$

In general, for 'X' to be consistent and the same at the exit of the reactors of both top and bottom streams, 'v' must be adjusted accordingly. That is,

$$\tau = \frac{V_1}{v_1} = \frac{V_2}{v_2} = \ldots$$

in all parallel breaks. In other words 'N' PFR in parallel will give the same conversion, as in a single PFR of volume $V = \sum V_i$ if $\tau_{total} = \tau_1 = \tau_2 = \ldots$.

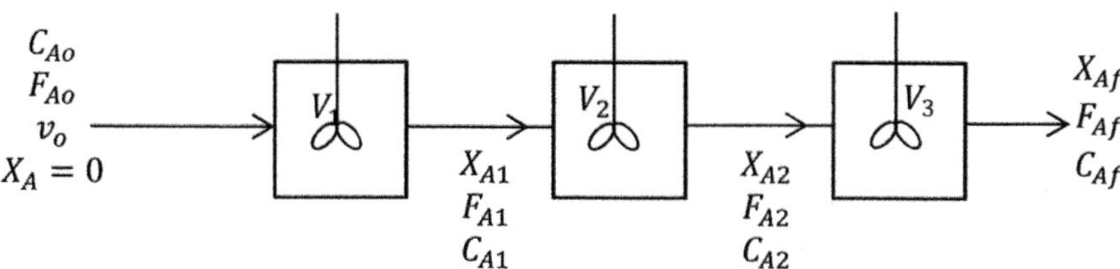

Recall

$$F_{A1} = F_{Ao}\left(1 - X_{A1}\right) \qquad F_{Af} = F_{Ao}\left(1 - X_{Af}\right)$$

$$C_{A1} = C_{Ao}\left(1 - X_{A1}\right) \qquad C_{Af} = C_{Ao}\left(1 - X_{Af}\right) \quad (\epsilon = 0)$$

(Conversion is defined with respect to unconverted stream).

Species balance (design equation):

$$\left.\begin{array}{l} F_{Ao} X_{A1} = (-r_A)_1\, V_1 \\ F_{Ao} X_{A2} \neq (-r_A)_2\, V_2 \\ \left(F_{Ao}\left(X_{A2} - X_{A1}\right) = (-r_A)_2\, V_2\right) \end{array}\right\} \begin{array}{l} F_{A1} = F_{Ao}\left(1 - X_{A1}\right) \\ F_{A2} = F_{Ao}\left(1 - X_{A2}\right) \qquad : moles/s \\ F_{A1} - F_{A2} = F_{Ao}\left(X_{A2} - X_{A1}\right) \end{array}$$

Design equations:

$$\text{reactor 2:} \qquad \tau_2 = \frac{V_2}{v_o} = \frac{C_{Ao}\left(X_{A2} - X_{A1}\right)}{(-r_A)_2}$$

$$\text{1:} \qquad \tau_1 = \frac{V_1}{v_o} = \frac{C_{Ao}\left(X_{A1}\right)_f}{(-r_A)_1}$$

$$\vdots$$

© The Author(s) 2025

N. Verma, *Chemical Reaction Engineering*,

https://doi.org/10.1007/978-3-031-88691-1_9

$$i: \quad \tau_i = \left(\frac{V}{v_o}\right)_i = \frac{C_{Ao}\,(X_{Ai} - X_{Ai-1})}{(-r_A)_i};$$

$$X_{Ai} = \frac{C_{Ai} - C_{Ao}}{C_{Ao}}$$

$$\tau_i = \frac{C_{i-1} - C_i}{kC_i} \quad \text{(for 1st-order)}$$

τ_i is space time or mean residence time ($\varepsilon_A = 0$).

$$\boxed{\frac{C_{i-1}}{C_i} = 1 + k\tau_i}$$

As τ_{is} are same in all branches

$$\frac{C_o}{C_N} = \frac{1}{1 - X_{AN}} = (1 + k\tau_i)^N$$

$$\tau_{N\text{-reactors}} = N\tau_i = \frac{N}{k}\left[\left(\frac{C_o}{C_N}\right)^{1/N} - 1\right]$$

(Same as before). In the limit; $N \to \infty$.

$$\tau_N = \frac{1}{k}\ln\frac{C_o}{C_N} = \tau \equiv \text{Plug flow;}$$

All it means is that the performance of a PFR is equivalent to that of infinite # of CSTR in series.

Consider four number of CSTR in series:

$$\frac{C_{Ao}}{C_{Af}} = \frac{1}{1 - X_{Af}} = (1 + k\tau_i)^4$$

$$\tau_i = \frac{V_1}{v_o} = \frac{\tau}{4}$$

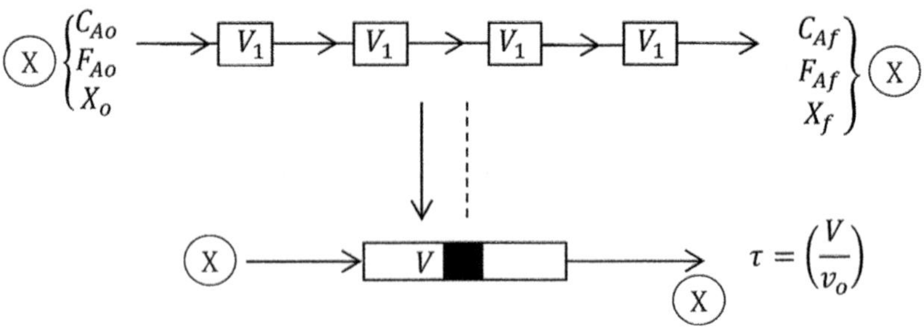

$$(X_{Af})_{PFR} > (X_{Af})_{CSTR} \quad \text{if } V = \Sigma V_1 = 4V_1$$

Similarly,

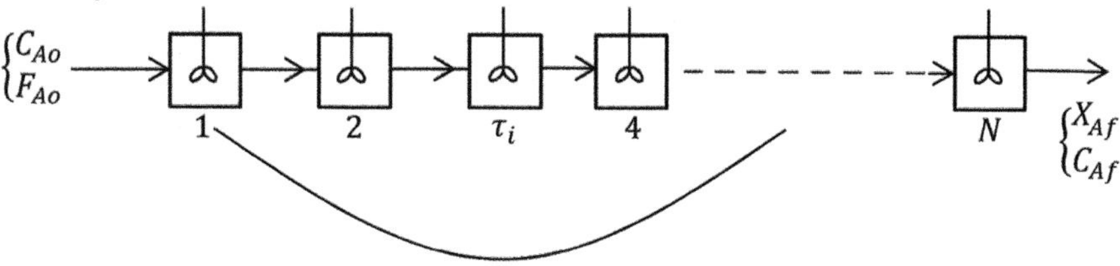

$$\sum_i^{N \to \infty} V_i \longrightarrow V_p \text{ for the same} C_{Af}(X_{Af})$$

Notes:

Performance of 'N' equal size CSTR^s: ($n > 0$)

(1) For the same $\tau \left(\frac{V}{v_o} \text{ and } \frac{\Sigma v_i}{v_o} \right)$, the latter will give higher conversion than the former. That is, for $V = 100L$ (say)

$$X \text{ from } \frac{100}{v_o} < X \text{ from } \frac{50 + 50}{v_o}$$

$$< X \text{ from } \frac{33.33 + 33.33 + 33.33}{v_o} < X \text{ from } \frac{100/(N-1)}{v_o}$$

$$< X \text{ from } \frac{100/N}{v_0}$$

$\quad = X$ from a PFR if $N \to \infty$ for the same v_o.

(2) For the same conversion and total volume V (let us say, 100L)

$$1 \times \tau_{100} > 2\tau_{50} > 3\tau_{33.3} > \cdots N\tau_{\frac{100}{N}} = \tau_p \ (N \to \infty)$$

or,

$$F_{Ao1} < F_{Ao2} < \cdots F_{Aon} = F_{Ao,p} \ (N \to \infty)$$

Graphical or numerical results for n equal size CSTRs in series are available for 1st-order:

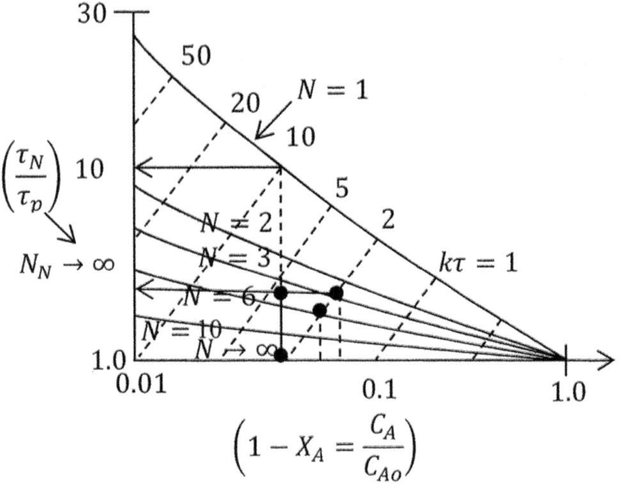

Procedure: $X_A, N \rightarrow k\tau$ (dotted lines)$\rightarrow V_{total}$ or v_0; y-axis gives additional information

$$\tau_N/\tau_p \left(\tau_p = \frac{1}{k} \ln \frac{C_o}{C} \right)$$

For second-order reaction graphical results are also available with $KC_{Ao}\tau$ as the parameter.

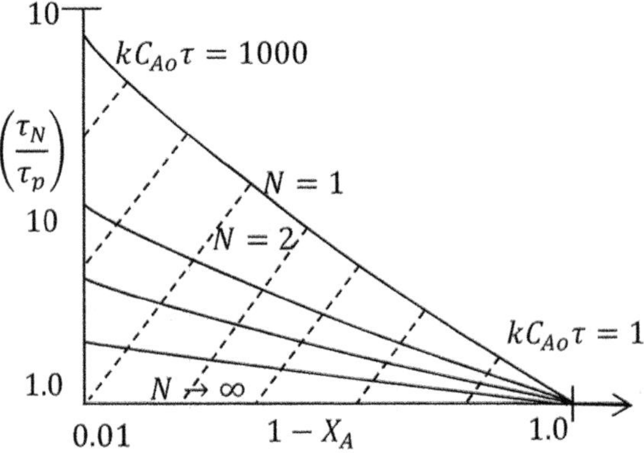

Mixed flow reactors of different sizes in series ($\epsilon_A = 0$)

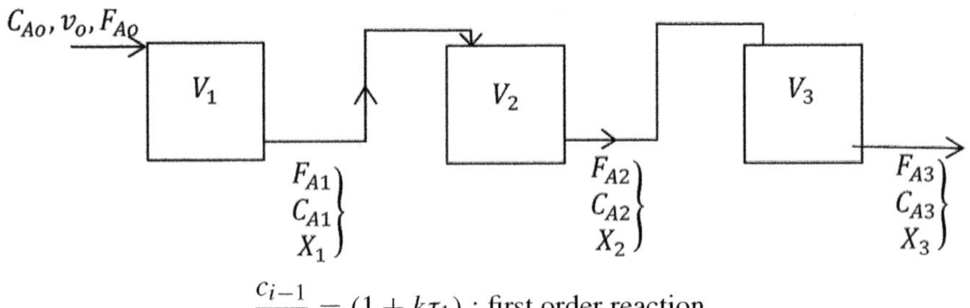

$$\frac{c_{i-1}}{c_i} = (1 + k\tau_1) : \text{first order reaction}$$

In general:

$$\tau_i = \frac{V_i}{v_0} = \frac{C_{Ao}\,(X_i - X_{i-1})}{(-r)_i}$$

or,

$$\tau_i = \frac{(C_{i-1} - C_i)}{(-r)_i}; \quad \left(X_i = \frac{C_o - C_i}{C_o} \right)$$

or,

$$-\frac{1}{\tau_i} = \frac{(-r)_i}{C_i - C_{i-1}} \qquad\qquad -\frac{1}{\tau_1} = \frac{(-r)_1}{C_1 - C_0}$$

$$-\frac{1}{\tau_2} = \frac{(-r)_2}{C_2 - C_1} \qquad\qquad -\frac{1}{\tau_3} = \frac{(-r)_3}{C_3 - C_2}$$

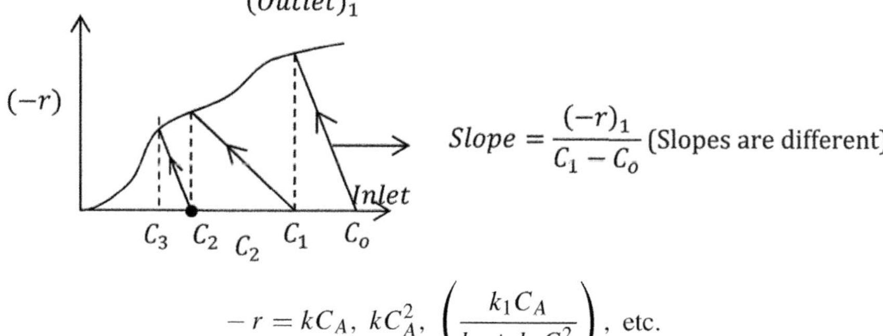

$$Slope = \frac{(-r)_1}{C_1 - C_o} \text{ (Slopes are different)}$$

$$-r = kC_A,\ kC_A^2,\ \left(\frac{k_1 C_A}{k_2 + k_3 C_A^2} \right),\ \text{etc.}$$

can be extended with $\varepsilon_A \neq 0$.

Reactors of different types in series:

In terms of design equations:

$$\tau_1 = C_{Ao} X_1 / (-r)_1;$$

$$\tau_2 = C_{Ao} \int_{X_1}^{X_2} \frac{dX}{(-r)_2};$$

$$\tau_3 = C_{Ao} \left(\frac{X_3 - X_2}{(-r)_3} \right)$$

$$(\tau_1 + \tau_2 + \tau_3) = \frac{V}{v_o} = C_{Ao}\left[\frac{X_1}{(-r)_1} + \int_{X_1}^{X_2}\frac{dX}{(-r)_2} + \left(\frac{X_3 - X_2}{(-r)_3}\right)\right]$$

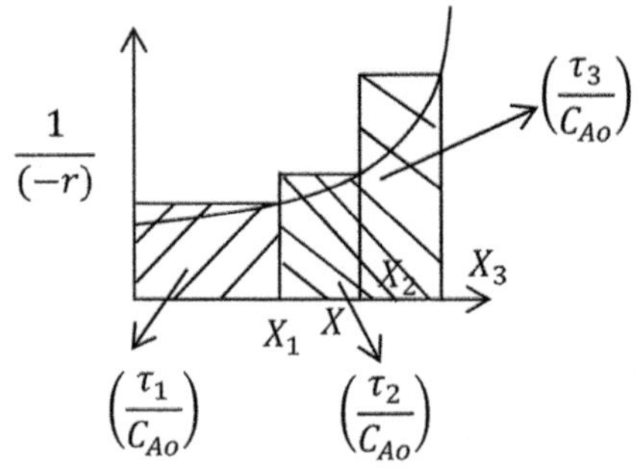

(In this chapter or elsewhere, consider 'X' as 'X_A', and $(-r)$ as $(-r_A)$)

Example 1

The following (isothermal) batch data are available for the liquid phase reaction
$A \rightarrow R$

C_A (moles/L)	0.1	0.2	0.3	0.4	0.6	0.8	1.2
$-r_A$ (moles/L-min)	0.1	0.3	0.5	0.6	0.25	0.06	0.045

(a) How long will it take for the concentration of A to decrease from 1.2 to 0.2 moles/L?

(b) The same reaction is to be carried out in a PFR of unknown volume followed by a MFR of 25 L volume connected in series. The specification of the feed stream is as follows: $v_o = 750$ L per h, $C_{Ao} = 1$ mole/L. If 80% conversion of A is required from the above PFR-MFR combination, what size of the PFR would be needed?

Ans:

(1) The first part of the question pertains to the batch reactor, but is also required to establish/determine batch kinetics or intrinsic rate, i.e., $1/(-r_A)$ vs C_A, which in turn is required for part (b) for the PFR-MFR system.

(2) How were $(-r_A)$ data, reported in part (a)? $(-r_A)$ must have been calculated from the time-concentration data:

t (min)		...	...	...	...
C_A (moles/L)		...	...	...	...
$(-r_A) = (\Delta C_A)/\Delta t$		...	...	...	...

(3) Calculate $\dfrac{1}{(-r_A)avg}$:

C_A	0.1	0.2	0.3	0.4	0.6	0.8	1.2
$(-r_A)$	0.1	0.3	0.5	0.6	0.25	0.06	0.045
$1/(-r_A)$	10.0	3.333	2.0	1.667	4.0	16.667	22.222
$1/(-r_A)$ avg		6.667	2.667	1.883	2.833	10.333	19.445

Batch reactor:

$$-\frac{dC_A}{dt} = (-r_A) \quad \text{(exact kinetics is not known)}$$

$$\Rightarrow - \int_{C_{Ao}=1.2}^{C_A=0.2} \frac{dC_A}{(-r_A)} = \int_0^t dt = t$$

$$t = \Sigma \frac{\Delta C_A}{(-r_A)\,avg}$$

$$= 0.1\,(2.667 + 1.833) + 0.2\,(2.833 + 10.333) + 0.4 \times 19.445$$

Choose $(-r_A)_{avg}$ over $0.2 - 1.2$ moles/lt $= 10.861$ min (See the plot below)

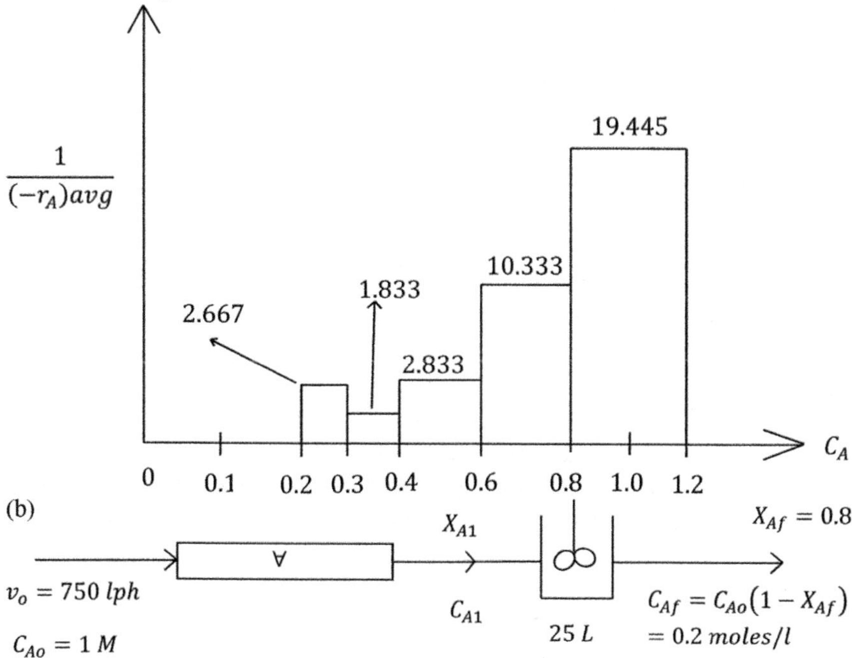

(b)

MFR:

$$\tau = \frac{V}{v_o} = \frac{25}{750/60} = 2 \text{ min} = C_{Ao}\frac{\Delta X_{Af}}{(-r_A)_f}$$

$$= \frac{1 \times (0.8 - X_{A1})}{0.3}$$

from batch data!

(kinetic rate at any concentration will be the same whether in a batch or MFR or PFR at the constant temperature)

$$X_{A1} = 0.2 \Rightarrow C_{A1} = C_{Ao}(1 - X_{A1}) = 0.8 \text{ mol/l}$$

PFR:

$$\tau = \frac{V}{v_o} = C_{Ao}\int_0^{0.2} \frac{dX_A}{(-r_A)}$$

$$= -\int_{1.0}^{0.8} \frac{dC_A}{(-r_A)} = \frac{\Delta C_A}{(-r_A)\,avg}$$

$$= (1.0 - 0.8) \times 19.445 \quad (\text{from batch data or plot})$$

$$V = v_o \times t$$

$$= \frac{750}{60} \times 0.2 \times 19.445$$

$$= 48.6 L$$

Note: Exact kinetics (order) was not known in this case; $(-r_A)$ was given against 'C_A'. Yet the data were sufficient to do calculation without knowing the exact expression for the rate.

Example 2

See below the arrangement of ideal CSTR and PFR in four branches. An aqueous phase reaction $A \rightarrow B$ $(k = 0.1\ s^{-1})$ is carried out under SS conditions in each reactor. Total volumetric flow rate of the feed ($C_{Ao} = 1\ M$) is 100 lpm. If the concentration of the reactant at the combined outlet (mixing point) branch is 0.5 M, determine V.

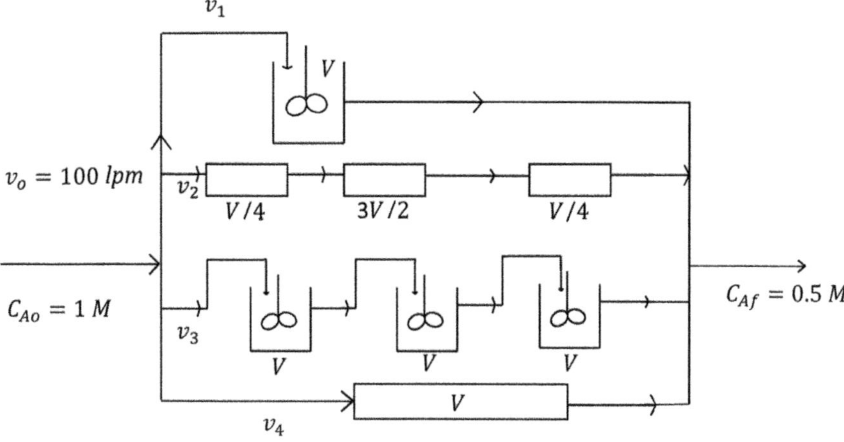

Ans:

(1)

$$v_o = 100 = v_1 + v_2 + v_3 + v_4 \text{ (flow rate)}$$

(2)

$$\tau_{PFR}\ (2-4\ \text{streams}) = \frac{\left(\frac{V}{4} + \frac{3V}{2} + \frac{V}{4}\right)}{v_2} = V/v_4$$

(τ must be the same in the parallel branch for PFR; $\tau_2 = \tau_4$)

$$v_2 = 2v_4 \quad \text{or} \quad v_4 = \frac{v_2}{2}$$

(3)

$$\tau_1 = \frac{V}{v_1} = C_{Ao}\frac{\Delta X_{Af}}{(-r_A)_f} = \frac{C_{Ao}\Delta X_{Af}}{kC_{Af}} = \frac{0.5}{k \times 0.5} = \frac{1}{k}; \quad v_1 = kV$$

(4)

$$\tau_2 = \frac{2V}{v_2} = C_{Ao}\int_0^{0.5} \frac{dX_A}{(-r_A)} = -\frac{1}{k}\ln(1 - X_{Af}), \quad v_2 = \frac{2kV}{0.693} = 2.886\,kV$$

(5)

$$\frac{C_{Ao}}{C_A} = \left(1 + k\frac{V}{v_3}\right)^3$$

$$1 + \frac{kV}{v_3} = \left(\frac{1}{0.5}\right)^{\frac{1}{3}} = 1.256, \quad v_3 = \frac{kV}{0.256} = 3.906\,kV$$

Note

$$\tau_3 = \frac{3V}{v_3} < 3\tau_1$$

(6) Substitute

$$100 = kV\left(+1 + 2.886 + 3.906 + \frac{2.886}{2}\right), \quad k = 0.1 \times 60 \text{ min}^{-1}$$

$$V = \frac{100}{0.1 \times 60 \times (1 + 2.886 + 3.906 + 1.443)} = 1.8\ L$$

Summary:

(1) $\tau_1 \neq \tau_2 \neq \tau_3$ for the same $C_{Ao} \to C_{Af}$

(2) $\tau_2 = \tau_4$ (for PFR in series or parallel for the same ΔC_A)

In other words ($\boldsymbol{n > 0}$)

- $V = \Sigma V_i$ for PFR in series or parallel for the same τ.
- $V = \Sigma V_i$ for MFR in parallel for the same τ, but
- $V > \Sigma V_i$ for MFR in series for the same τ.

Example 3

Pure gaseous reactant A ($C_{Ao} = 100$ mM/L is fed at a flow rate of 1.5 LPH into a well mixed reaction ($V = 0.1\ L$) at the constant temperature of 150°C, where it dimerizes by the elementary reaction $2A \to A_2$ under the SS condition. The concentration of A is measured as 33.4 mM/L. During operation on a certain day, the inlet and outlet valves of the reactor malfunction, completely shutting off the inlet and outlet gas stream to and from the reactor, respectively. The reactor temperature, however, remains constant at 150°C. It will take one hour for the plant technician to fix the problem and restart the gas flow. What concentration of A will be measured in the reactor at the instant of restarting the flow in the reactor?

Ans:

$$C_{Ao} = 100 \text{ mM/L} \qquad \boxed{V = 0.1\ L} \qquad C_{Af} = 33.4 \text{ mM/L}$$

$$MFR$$

$$v_o = 15 \text{ LPH} \qquad -r_A = kC_A^2$$

Species balance:

$$v_o C_{Ao} - v_f C_{Af} = V(-r_A)_f = V(kC_{Af}^2)$$

$$2A \to A_2; \quad \varepsilon_A = \frac{\frac{1}{2} - 1}{1} = -0.5$$

$$v_f = v_o \left(1 + \varepsilon_A X_{Af}\right) = v_o(1 - 0.5 X_{Af})$$

$$\implies C_{Af} = \frac{C_{Ao}\left(1 - X_{Af}\right)}{\left(1 + \varepsilon_A X_{Af}\right)} \implies 33.4 = \frac{100\left(1 - X_{Af}\right)}{\left(1 - 0.5 X_{Af}\right)}$$

$$X_{Af} = 0.80$$

Therefore,

$$100 \times 1.5 - 1.5\,(1 - 0.5 \times 0.8) \times 33.4 = 0.1\left(k \times 33.4^2\right)$$
$$\implies k = 1.075 \text{ L/h-mM}$$

Alternatively, the **design equation**

$$\tau = \frac{V}{v_o} = \frac{C_{Ao} X_{Af}}{(-r_A)_f}$$

will also work. Design equation is nothing but the species balance!!

Batch reactor (constant volume):
When the power goes off for 1 h but the temperature remains the same $\implies k = 1.075$ L/h-mM. The rate constant depends on temperature and not on the type of reactor (flow or batch).

$$- r_A = -\frac{1}{V}\frac{dN_A}{dt} = -\frac{dC_A}{dt} = kC_A^2 \quad (V = \text{constant})$$

$$\frac{1}{C_A} - \frac{1}{C_{Ao}} = kt \implies \frac{1}{C_A} - \frac{1}{33.4} = 1.075 \times 1$$

$$C_A = 0.905 \text{ mM/L}$$

Example 4
At present 75% of the reactant A in the liquid phase is converted to product by a 1st-order reaction in a single MFR

(a) We plan to place two additional reactors similar to the one being used in series with it.

    (i) For the same treatment rate as that used at present, how will this addition affect the conversion of the reactant?

    (ii) For the same 75% conversion, by how much can the treatment rate be increased in the new configuration?

(b) If the both additional two reactors are to be operated in parallel with the first single reactor, i.e., all three reactors in parallel, by how much can treatment be increased for the same 75% conversion?

(c) If the additional two reactors are placed in series but in parallel with the first single reactor, by how much can the treatment be increased for the same 75% conversion?

(d) Using any number of equal size CSTR or MFR in series, by how much maximum can the total volume be reduced for 75% conversion, if the treatment rate is the same as that used in part (i) of (a)?

Ans: Single MFR:

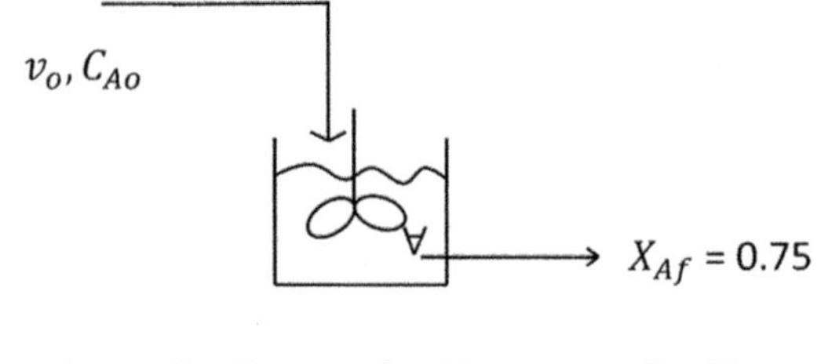

$$\tau = \frac{V}{v_o} = \frac{C_{Ao}X_{Af}}{(-r_A)_f} = \frac{C_{Ao}X_{Af}}{kC_{Af}} = \frac{C_{Ao}X_{Af}}{kC_{Ao}(1 - X_{Af})}$$

$$k\tau = \frac{X_{Af}}{1 - X_{Af}} = \frac{0.75}{(1 - 0.75)} = 3$$

(a)

(i)

$$\frac{C_{Ao}}{C_A} = (1 + k\tau_i)^N = (1 + 3)^3 = 64$$

$$\frac{C_{Ao}}{C_{Ao}(1 - X_{Af})} = 64 \implies X_{Af} = 0.98$$

(ii)

$$\frac{C_{Ao}}{C_A} = (1 + k\tau_i)^N$$

$$\frac{C_{Ao}}{C_{Ao}(1 - X_{Af})} = (1 + k\tau_i')^3 \implies 1 + k\tau_i' = \left(\frac{1}{1 - 0.75}\right)^{\frac{1}{3}}$$

$$k\tau_i' = 0.586 \implies \frac{\tau_i'}{\tau_i} = \frac{0.586}{3}$$

or

$$\frac{v_o'}{v_o} = \frac{3}{0.586} = 5.10 \quad \left(\tau = \frac{V}{v_o}\right)$$

(b)

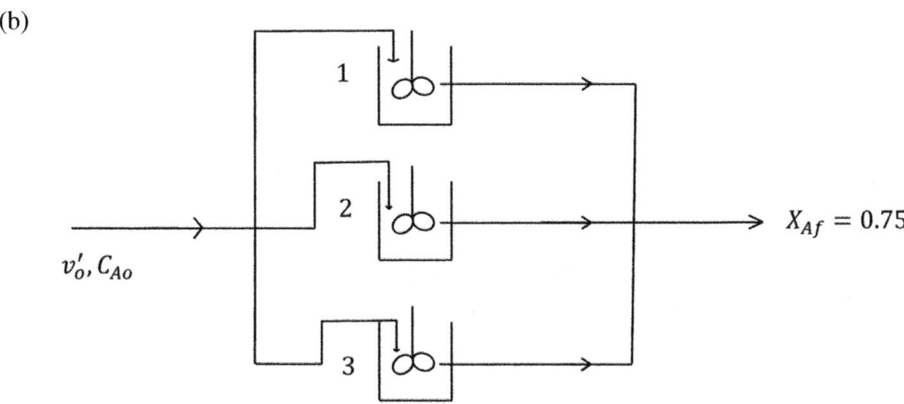

$$\tau_i' = \frac{V}{v_o'/3}$$

(maintain the same τ in all '3' parallel-reactors i.e., flow rate is split-equally). Therefore, for the same conversion (75%), $v_o' = 3v_o$ (3-times) or $\tau_i' = \tau_i$.

(c)

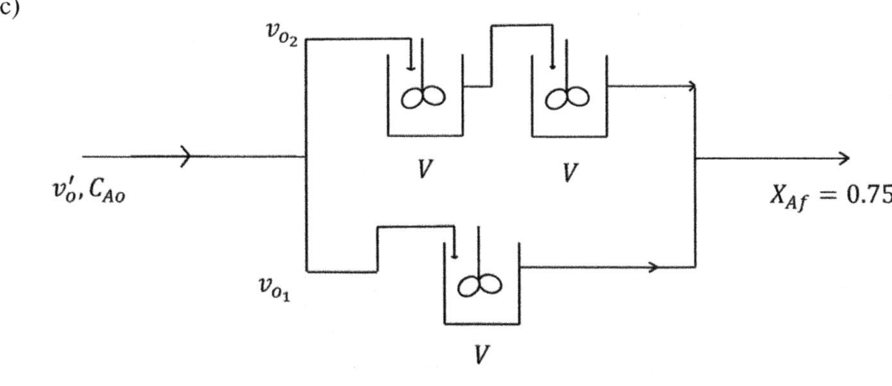

$$v'_o = v_{o2} + v_{o1}$$

The exit concentration of both branches must be the same i.e.,

$$X_{Af} = 0.75$$

1st-branch:

$$\frac{C_{Ao}}{C_A} = (1 + k\tau_i)^2$$

$$k\tau_{i,2} = \left(\frac{1}{1-0.75}\right)^{\frac{1}{2}} - 1 = 1 \quad (C_A = C_{Ao}(1-0.75))$$

2nd-branch:

$$\frac{C_{Ao}}{C_A} = 1 + k\tau_{i,1}$$

$$k\tau_{i,1} = \frac{1}{1-0.75} - 1 = 3$$

$$v_{o2} + v_{o1} = kV + k\frac{V}{3} = \frac{4kV}{3}$$

$$k\tau = \frac{kV}{v_o} = 3: \quad \text{Single reactor (original case)}$$

$$v_o = \frac{kV}{3}$$

$$\frac{v_{o2} + v_{o1}}{v_o} = 4 \text{ (four-times)}$$

(d) Any number of equal size CSTRs giving the smallest volume implies using a PFR.

$$k\tau_{PFR} = -\ln \frac{C_A}{C_{Ao}} \quad \text{(PFR design equation)}$$

$$k\tau_{PFR} = -\ln(1 - X_A) = -\ln(1 - 0.75) = 1.386$$

$$\frac{\tau_{PFR}}{\tau_{CSTR}} = \frac{1.386}{3} \implies \frac{V_{PFR}}{V_{CSTR}} = 0.46 \text{ (same } v_o)$$

Example 5

A homogeneous gas reaction $A \to 2B$ is run at 100°C at a constant pressure of 1 atm in an experimental batch reactor. The data in table below were obtained starting with pure A. What size of PFR operated at 100°C and 10 atm would yield 90% conversion of A for a total feed rate of 10 mol/s, the feed containing 40% inerts?

Ans:

(1) Order of the reaction is not known; stoichiometry is known.
(2) You may like to recall. $\tau_{PFR} = t_{batch}$ only if $\varepsilon_A = 0$, which is not the case here.

$$\tau_{PFR} = \frac{V}{v_o} = C_{Ao} \int_0^{X_A} \frac{dX_A}{(-r_A)} \text{ for any } \varepsilon_A$$

$$t_{batch} = C_{Ao} \int_0^{X_A} \frac{dX_A}{(-r_A)} \quad (V = C)$$

$$(F_{Ao} dX_A = (-r_A) \, d\forall)$$

$Time, min$	0	1	3	5	7	9	11	13	14
V/V_o	1.0	1.2	1.48	1.66	1.78	1.86	1.91	1.94	1.95

Hint: Check if it pertains to zero or 1st or any other order–kinetic. Determine k.

Example 6

The data in the table below have been obtained on the decomposition of gaseous reactant A in a constant volume batch reactor at 100°C. The stoichiometry of the reaction is $2A \rightarrow R + S$. What size of a PFR in liters operating at 100°C and 1 atm can treat 100 mol of A/hour in a feed consisting of 20% inerts to obtain 96% conversion of A?

$t(s)$	$p_A(atm)$	$t(s)$	$p_A(atm)$
0	1.0	140	0.25
20	0.8	200	0.14
40	0.68	260	0.08
60	0.56	330	0.04
80	0.45	420	0.02
100	0.37		

Ans:

(1) The reaction $2A \rightarrow R + S$ carried out in a constant volume reactor will not show any change in total pressure. How was p_A (atm) then reported? It must have been calculated from $C_A RT$, with 'C_A' measured using some chemical analysis or sampling; else you can use the expression:

$$p_A = p_{Ao} - \frac{a}{\Delta n}(P - P_o)$$

without any sampling or chemical analysis.

(2) Recall if $\varepsilon_A = 0$ like here, $\tau_{\text{PFR}} = t_{\text{batch}}$

$$\tau_{PFR} = \frac{V}{v_o} \equiv C_{Ao} \int_0^{X_A} \frac{dX_A}{(-r_A)}$$

$$t_{batch} = C_{Ao} \int_0^{X_A} \frac{dX_A}{(-r_A)} \quad (\varepsilon_A = 0)$$

$$\tau_{PFR} = t_{batch} \quad \text{if} \, C_{Ao} \text{of batch} = C_{A,inlet} \text{ of PFR}$$

$$C_{A,inlet} = \frac{p_A}{RT} = \frac{0.8 \times 1}{0.082 \times 373} = 0.026 \text{ moles/L}$$

Prepare the table of batch-data using X_A with respect to $C_A = \underline{0.026}$, otherwise you have made a mistake.

$t(s)$	0	$\underline{20}$	...	$\underline{330}$	420
$p_A(atm)$	1.0	0.8	...	0.04	0.02
$C_A = \dfrac{p_A}{RT}$ $(moles/L)$	0.032	$\underline{0.026}$	...	$\underline{0.0013}$	0.000654
$\left(X_A = \dfrac{C_{Ao} - C_A}{C_{Ao}}\right)$	0	0.1875	...	0.959	0.979 **mistake**
X_A	–	$\underline{0}$	...	$\underline{0.96}$	**correct**

$$t_{batch} = (330 - 20) = 310 \, s \, \text{for} \, 96\% \text{conversion}$$

$$\tau_{PFR} = 310 = \frac{V}{v_o} = \frac{V C_{Ao}}{F_{Ao}}$$

$$V = 310 \times \frac{F_{Ao}}{C_{Ao}} = 310 \times \left(\frac{100}{60 \times 60 \times 0.026}\right) = \underline{331.20 \, L}$$

(Be careful in working or 'X_A'. It is an auxiliary variable. For batch reactor, it is defined with respect to initial concentration, and for flow-reactors it is defined with respect to inlet concentration. They were different here. An important thing also to note is that neither $(-r_A)$ or nor rate is required to be calculated in this case. The integral is also not required to be calculated).

Example 7

The homogeneous gas reaction $A \rightarrow 3R$ follows the 2nd-order kinetics. For a feed rate of 4 m³/hr of pure A at 5 atm and 350°C, an experimental reactor consisting of a 2.5 cm 1D pipe 2 m long gives 60% conversion of feed. A commercial plant is to treat 320 m³/hr of feed consisting of 50% A, 50% inerts at 350°C and 25 atm to obtain 80% conversion.

How many 2 m lengths of 2.5 cm 1D pipe are required?

Ans: This is an example on using the data for two PFRs having different feed conditions.

$$-r_A = -\frac{dC_A}{dt} = kC_A^2 \qquad A \rightarrow 3R; \quad \varepsilon_A = 2$$

Case 1:

$$V_{PFR} = \left(\frac{\pi}{4}d^2\right) L; \quad v_o = 4\,m^3/h; \quad X_{Af} = 0.6 \ (d = 2.5\,cm,\ L = 2\,m)$$

$$V = v_o C_{Ao} \int_0^{X_{Af}} \frac{dX_A}{(-r_A)}$$

$$C_A = \frac{C_{Ao}(1 - X_A)}{(1 + \varepsilon_A X_A)}$$

$$C_{Ao} = \frac{p_{Ao}}{RT} = \frac{y_{Ao}p}{RT} = \frac{5}{R \times (350 + 273)}$$

$$V = 4 \times C_{Ao} \times \int_0^{0.6} \frac{(1 + \varepsilon_A X_A)^2}{kC_{Ao}^2(1 - X_A)^2} dX$$

$$= \frac{4R \times 623}{5k} \int_0^{0.6} \frac{(1 + 2X_A)^2}{(1 - X_A)^2} dX$$

Case 2:

$$v_o = 320\,\frac{m^3}{h}, \quad X_{Af} = 0.8, \quad \varepsilon_A = \frac{(3 \times 0.5 + 0.5) - 1}{1} = 1, \ T = 623\,K$$

$$C_{Ao} = \frac{p_{Ao}}{RT} = \frac{y_{Ao}p}{RT} = \frac{0.5 \times 25}{R \times 623}$$

$$V = \frac{320R \times 623}{0.5 \times 25k} \int_0^{0.8} \frac{(1 + X_A)^2}{(1 - X_A)^2} dX$$

$$\frac{V_2}{V_1} = \frac{320 \times 5}{0.5 \times 25 \times 4} \frac{\int_0^{0.8} \frac{(1 + X_A)^2}{(1 - X_A)^2} dX}{\int_0^{0.6} \frac{(1 + 2X_A)^2}{(1 - X_A)^2} dX} = 32 \times \text{Integral} = 67.6$$

(Integral is complex but not difficult to evaluate! Check)

$$V_2 = 67.6 \; V_1 = 67.6 \times \left(\frac{\pi}{4} \times (2.5)^2 \times 200 \right) \; cm^3$$

We require $\sim \underline{68}$ number of 2 m length of 2.5 cm ID pipe.

Example 8

Consider the following batch kinetic data of a liquid-phase chemical reaction $A \rightarrow B$.

X_A	0	0.1	0.2	0.3	0.4	0.5	0.6	0.7	0.8
$\left(\dfrac{C_{Ao}}{(-r_A)} \right)$, min	5	25	40	45	35	20	10	6	10

(a) Which of the two arrangements are preferable if the total volume requirement is to be minimum? A MFR followed by PFR or PFR followed by MFR? In either case, the final conversion is 0.7 and the intermediate conversion between the two reactors is 0.3. Explain in terms of the kinetic rate.

(b) For the chosen configuration in (a), what is the total volume required to obtain $X_A = 0.7$ at the volumetric flow rate of feed $= 5 \; lpm$?

(c) Given all options for the choice of reactor and configuration, what is the minimum volume required to obtain $X_A = 0.7$ at the volumetric flow rate of the feed $= 5 \; lpm$?

(d) What is the required volume of a single PFR operating at the average rate of 0.04 $\frac{mol}{min-L}$ to obtain $X_A = 0.7$? Feed: Flow rate$= 50$ lpm, $C_{Ao} = 1M$.

(e) If a single PFR ($ID = 20$ cm, $l = 10$ m) is used to treat the feed of part (d) with the exit conversion $X_A = 0.7$, determine the concentration of A at the distance $l/10$ from the inlet and $l/10$ from the outlet of the reactor?

Ans: Plot (qualitatively, but accurately)

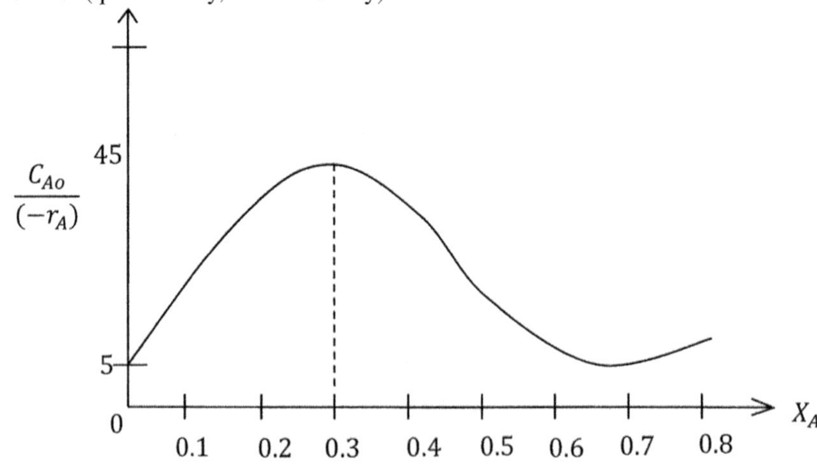

$\Rightarrow (-r_A)$ decreases with increasing 'X_A' till $X_A = 0.3$: +ve order reaction: use PFR.

$\Longrightarrow$ It increases with increasing 'X_A' from 0.3 to 0.7 $-$ve order reaction: use MFR.

(a) PFR followed by MFR for the **maximum rate** in the reactors or minimum volume.
(b)

$$V_{total} = V_{PFR} + V_{MFR}$$

$$\tau_{PFR} = \frac{V}{v_o} = C_{Ao} \int_0^{0.3} \frac{dX_A}{(-r_A)}$$

$$\tau_{MFR} = \frac{V}{v_o} = \frac{C_{Ao}\Delta X_A}{(-r_A)_f}$$

$$V_{PFR} = 5.0 \times \left[\frac{5+25}{2} + \frac{25+40}{2} + \frac{40+45}{2} \right] \times 0.1 : \left(\frac{f_1+f_2}{2} \Delta X_A \right)$$

$-$Trapezoidal rule of integration: $\Delta X_A = 0.1$

$$V_{MFR} = 5.0 \times (0.7 - 0.3) \times 6 \quad \left(\text{Note: } \frac{C_{Ao}}{(-r_A)_f} = 6 \text{ at } X_A = 0.7 \right)$$

$$V_{total} = 57\ L$$

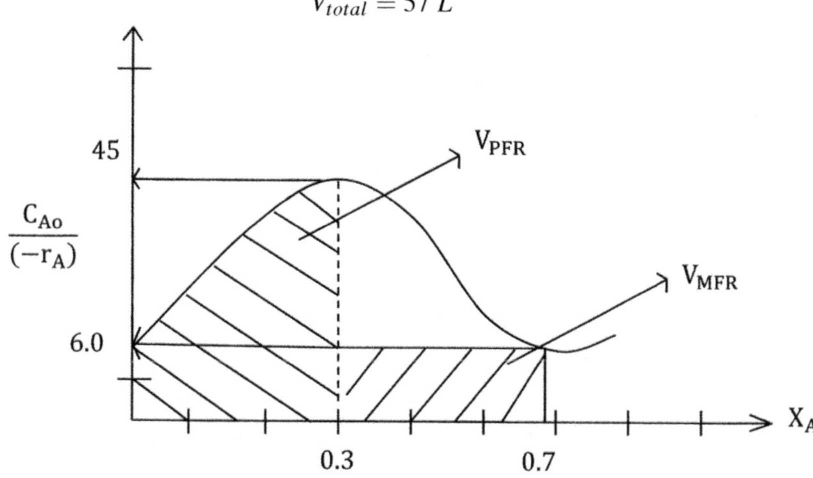

(c) V_{min} or τ_{min} will come from a single MFR operating at

$$X_{Af} = 0.7 \Longrightarrow \frac{C_{Ao}}{(-r_A)} = 6.0$$

$$\tau = \frac{V}{v_o} = C_{Ao}\frac{\Delta X_{Af}}{(-r_A)_f} \Longrightarrow V = 5 \times (0.7 - 0) \times 6 = 21\ L$$

(d)

$$\tau_{PFR} = C_{Ao} \int_0^{X_{Af}} \frac{dX_A}{(-r_A)} = C_{Ao} \underbrace{\frac{X_{Af}}{(-r_A)}}_{\text{average rate}}$$

$$V_{PFR} = 50 \times \frac{1 \times 0.7}{0.04} = 875 \ L$$

(e)

$$F_{Ao} dX_A = (-r_A) dV$$

$$= (-r_A) A_c dz \quad \left(A_c = \frac{\pi}{4} d^2 \right)$$

$$\frac{dX_A}{dz} = \left(\frac{-r_A}{C_{Ao}} \right) \frac{A_c}{v_o} = \left(\frac{-r_A}{C_{Ao}} \right) \times \frac{0.034}{50 \times 10^{-3}} \ m^{-1}$$

$$\left(A_c = \frac{\pi}{4} d^2 = \frac{\pi}{4}(0.2)^2 = 0.034 \ m^2 \right)$$

$$\boxed{\frac{dX_A}{dz} = 0 + 0.6285 \left(\frac{-r_A}{C_{Ao}} \right); \quad z = 0, \ X_A = 0}$$

$$X_A|_{z=1 \ m} = 0 + 0.6285 \left(\frac{-r_A}{C_{Ao}} \right) \times 1 \quad \left(\text{check plot } \frac{(-r_A)}{C_{Ao}} = \frac{1}{5} = 0.2 \right)$$

$$\left. \begin{array}{c} z = 0 \\ X_A = 0 \end{array} \right\} \ \left(\text{Forward Euler}, \Delta z (\text{step-size is small}: \frac{l}{10} = 1 \ m) \right)$$

$$= 0.1257; \ C_A = C_{Ao}(1 - X_A) = \underline{0.874 M}$$

$X_A|_{z=1 \ m}$ from the exit or end $(X_{Af} = 0.7)$ of the tube.

$$\frac{dX_A}{dz} = 0.6285 \times \frac{1}{6} = 0.1043 \quad \left(\frac{(-r_A)}{C_{Ao}} = \frac{1}{6} \right)$$

$$X_A = 0.7 - 0.1043 \times 1 = 0.6 \ (\text{this is also Forward Euler})$$

$$C_A = \underline{0.04 \ M}$$

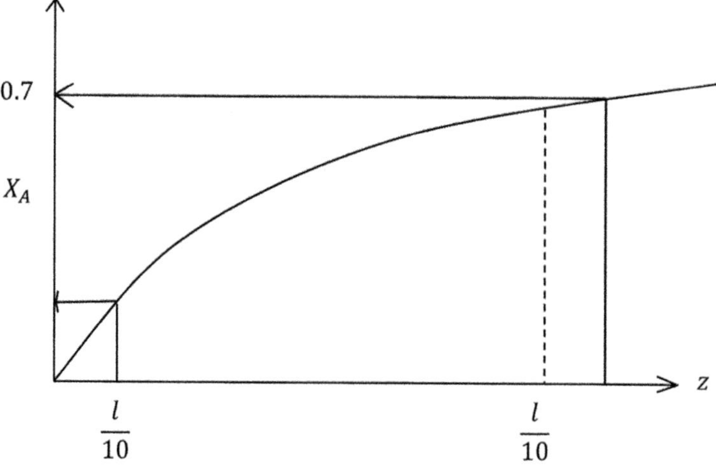

Advantageous in processing unconverted products, increasing mixing, creating temperature uniformity.

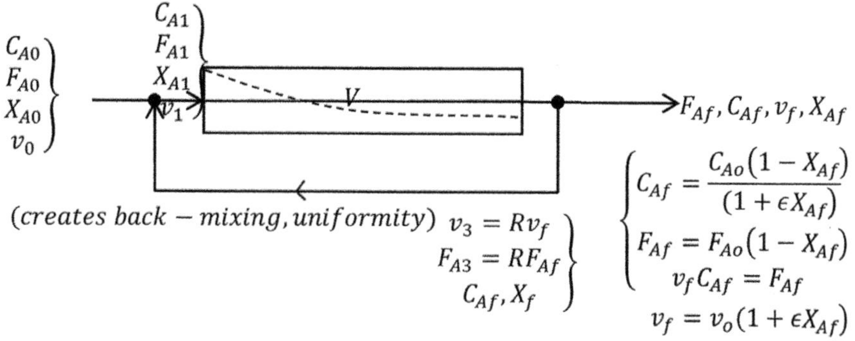

$$F_{A1} = F_{Ao} + F_{A3} = \left(F_{Ao} + RF_{Af}\right)$$

$$v_1 = v_o + v_3 = \left(v_o + Rv_f\right)$$

$$v_1 C_{A1} = F_{A1} \Rightarrow C_{A1} = \frac{F_{Ao} + RF_{Af}}{v_o + Rv_f} \qquad (1)$$

$$C_{A1} = \frac{C_{Ao}(1 - X_{A1})}{(1 + \epsilon X_{A1})} : \text{definition} \qquad (2)$$

Problem Definition: (F_{Ao}, C_{Ao}, v_o) are given to achieve (X_{Af}, C_{Af}, F_{Af}), and recycle rate is 'R'. What would be V?

Note: R is an additional variable in this case.

© The Author(s) 2025

N. Verma, *Chemical Reaction Engineering*,

https://doi.org/10.1007/978-3-031-88691-1_10

Performance Equation (without recycle):

$$\left.\begin{aligned}
F_{Ao}dX &= C_{Ao}(-r_A)dV \\
\tau_p = \frac{V}{v_o} &= C_{Ao}\int_0^{X_{Af}} \frac{dX}{(-r_A)}
\end{aligned}\right\}$$

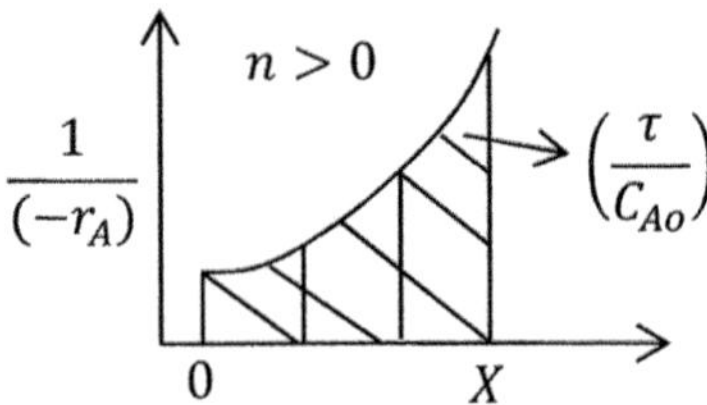

with recycle:

$$F'_{Ao}dX = (-r_A)dV$$

$$\tau_R = \frac{V}{v'_0} = C_{Ao}\int_{X_{A1}}^{X_{Af}} \frac{dX}{(-r_A)}$$

where

$$F'_{Ao} = F_{Ao} + RF_{Ao}$$

or

$$v'_o = v_o + Rv_o$$

$X_{A1} = ? \rightarrow$ Intermediate variable.

From (1) and (2)

$$\begin{aligned}
C_{A1} &= \frac{C_{Ao}(1 - X_{A1})}{(1 + \epsilon_A X_{Af})} \\
&= \frac{F_{Ao} + RF_{Af}}{v_o + Rv_f} \\
C_{A1} &= \frac{F_{Ao} + RF_{Ao}(1 - X_{Af})}{v_o + Rv_o(1 + \epsilon_A X_{Af})} \\
&= C_{Ao}\frac{1 + R - RX_{Af}}{1 + R + R\epsilon_A X_{Af}}
\end{aligned}$$

$$X_{A1} = \left(\frac{R}{R+1}\right)X_{Af}$$

Therefore,

$$\tau_R = \frac{V}{v'_0} = C_{Ao}\int_{\left(\frac{R}{R+1}\right)X_{Af}}^{X_{Af}} \frac{dX}{(-r_A)}$$

or

$$\tau = \frac{V}{v_0} = (R+1)C_{Ao}\int_{\left(\frac{R}{R+1}\right)X_{Af}}^{X_f}\frac{dX}{(-r_A)}$$

Important

$$\begin{cases} V = C_{Ao}v_o(R+1)\int_{\left(\frac{R}{R+1}\right)X_{Af}}^{X_{Af}}\frac{dX}{(-r_A)} \\[2em] \quad = F_{Ao}(R+1)\int_{\left(\frac{R}{R+1}\right)X_{Af}}^{X_{Af}}\frac{dX}{(-r_A)} \end{cases}$$

(Note the difference between τ_R & τ)

Inspect:

$$R = 0, \quad \tau = \frac{V}{v_o} = C_{Ao}\int_{0}^{X_{Af}}\frac{dX}{(-r_A)}: \quad \text{PFR}$$

$$R \to \infty, \quad \tau = C_{Ao}\frac{X_{Af}}{(-r_A)_f}: \quad \text{CSTR}$$

$$\tau_R = \frac{V}{v_o'=(R+1)v_o} = C_{Ao}\int_{X_{A1}=\left(\frac{R}{R+1}\right)X_{Af}}^{X_{Af}}\frac{dX_A}{(-r_A)}$$

$$R=0 \qquad\qquad\qquad\qquad\qquad\qquad R=\infty$$

$$\tau_p = \frac{V}{v_o} = C_{Ao}\int_{0}^{X_{Af}}\frac{dX}{(-r_A)} \qquad\qquad \tau_m = \frac{V}{v_o} = C_{Ao}\frac{X_{Af}}{(-r_A)_f}$$

Plug flow $\qquad\qquad\qquad\qquad\qquad$ *mixed flow*

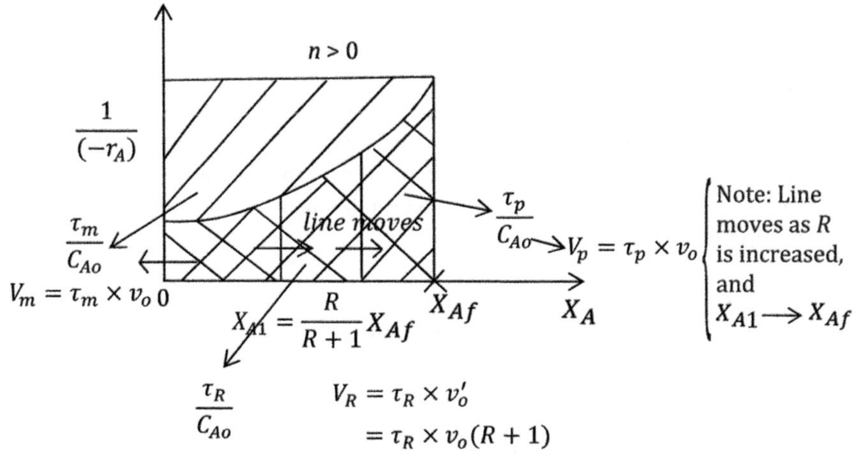

$$\begin{cases} R \longrightarrow \infty & V_R \longrightarrow V_m \\ R \longrightarrow 0 & V_R \longrightarrow V_p \end{cases}$$

Graphical Results:

$$\text{PFR} \quad 0 \longleftarrow R \longrightarrow \infty \quad \text{CSTR}$$

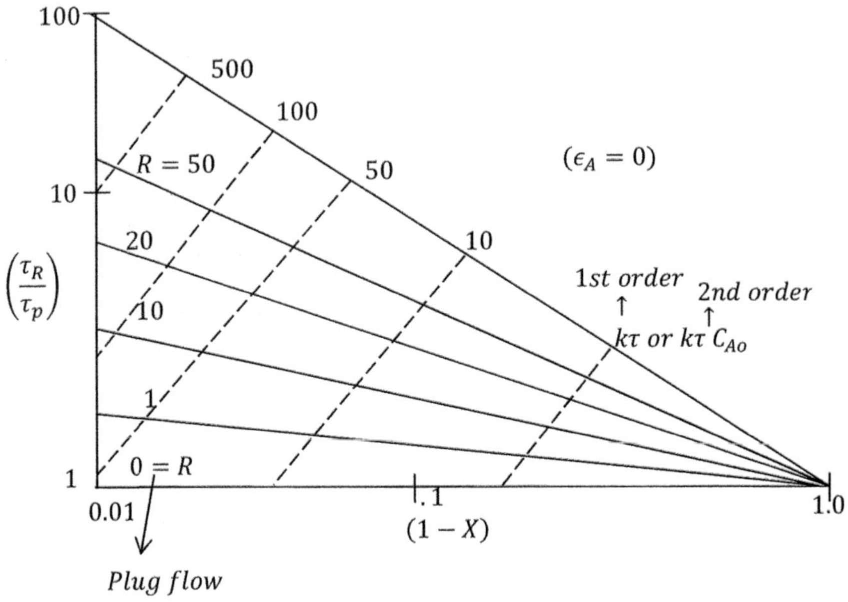

Example 1

The non-elementary, irreversible, aqueous phase reaction

$$A + B \rightarrow R \; (-r_A = k C_A)$$

is carried out isothermally under SS condition, as follows. The equal volumetric flow rates of two liquid streams are introduced into a 4 L mixing tank. One stream contains 0.02 moles/L of A and the other contains 1.4 moles/L of B. The mixed stream from the tank is passed through a recycle-tubular reactor of 12 L volume. We find that some R is formed at 0.002 moles/L concentration in the mixing tank. Assuming that the mixing tank acts as an ideal MFR, determine the concentration of R at the exit of the tubular reactor if

(a) No recycling is used for the tubular reactor.
(b) There is a recycle of the exit stream from the tubular reactor at the same flow rate as before (i.e., w/o recycle).

Ans: Firstly, you should draw the schematics of the arrangement and note down the given data as much as possible.

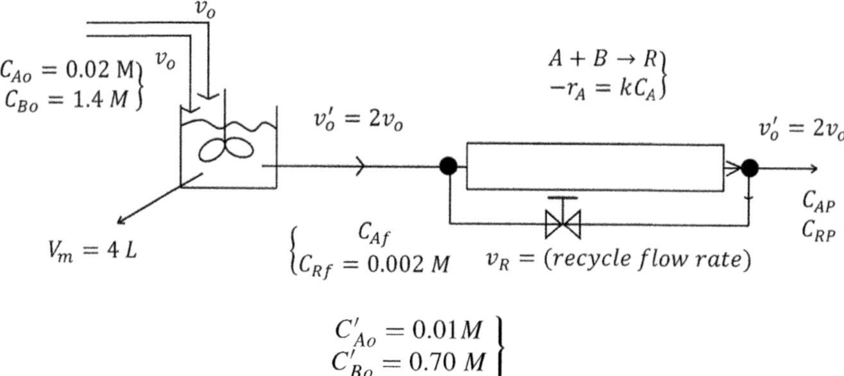

$$C'_{Ao} = 0.01\,M \atop C'_{Bo} = 0.70\,M$$

There is the effect of dilution on mixing. These are the actual inlet concentrations to MFR. Note the flow rates of the individual stream is v_o. Therefore, the concentration of both streams is practically or effectively halved

$$v_o' = v_o + v_o = 2v_o$$

Make use of the stoichiometry:

$$C_{Rf} = 0.002M$$
$$C_{Af} = 0.01 - 0.002 = 0.008M$$
$$\tau_m = \frac{C_{Ao} - C_{Af}}{(-r_A)_f} = \frac{0.01 - 0.008}{k(0.008)} = \frac{1}{4k} \quad (-r_A = kC_A)$$

$$v_o' = V_m \times 4k = 16k \ \left(\tau_m = \frac{V_m}{v_o'}\right)$$

Case a: Without recycle (valve in the recycle stream is closed)

$$\tau_p = \frac{V_p}{v_o'} = -\int_{C_{Af}}^{C_{AP}} \frac{dC_A}{kC_A} = \frac{1}{k}\ln\left(\frac{C_{Af}}{C_{AP}}\right)$$

$$\Longrightarrow \ln\left(\frac{0.008}{C_{Ap}}\right) = \frac{k \times 12}{16\,k} = \frac{3}{4} \Longrightarrow C_{Ap} = 0.0038\ M$$

$$C_{RP} = \left(C_{Af} - C_{AP}\right) + C_{Rf} \Longrightarrow \left(C_{Af} + C_{Rf} = C_{RP} + C_{AP}\right)$$
$$= (0.008 - 0.0038) + 0.002 = \underline{0.0062\ M}$$

$$v_R = v_o' = 2v_o$$

$$R = \frac{v_R}{\text{exit flow rate from the reactor-system}} \text{(definition)}$$

$$= \frac{2v_o}{2v_o} = 1 \text{ (Note this is SS condition)}$$

Recycle flow rate is the same as the exit flow rate; the same as the inlet flow rate; i.e., $v_o' = 2v_o$. Recall tubular reactor with recycle. Pay special attention to the schematic;

$$\tau_p = \frac{V}{v_o'} = (R+1)\, C_{Af} \int_{\left(\frac{R}{R+1}\right)X_{AP}}^{X_{AP}} \frac{dX_A}{(-r_A)} \quad (-r_A = kC_A)$$

$C_{Af} \equiv$ Inlet concetation to recycle reactor system (before mixing) $= 0.008$

$$X_{A1} = \left(\frac{R}{R+1}\right)X_{AP} = \frac{1}{2}X_{AP}$$

$$\tau_p = \frac{V_p}{v_o'} = (R+1)\, C_{Af} \int_{X_{\frac{AP}{2}}}^{X_{AP}} \frac{dX_A}{kC_{Af}(1-X_A)}$$

$$\frac{12}{16k} = \frac{2}{k}\ln\left(\frac{1-X_{\frac{AP}{2}}}{X_{AP}}\right), \quad (v_o' = 16k)$$

$$\ln\left(\frac{1-X_{\frac{AP}{2}}}{X_{AP}}\right) = \frac{3}{8} \implies \frac{1-X_{\frac{AP}{2}}}{X_{AP}} = 1.455; \quad X_{AP} = 0.511$$

$$C_{AP} = C_{Af}(1-X_{AP}) = 0.008 \times (1 - 0.511) = \underline{0.0039}$$

$$C_{RP} = (C_{Af} - C_{AP}) + C_{Rf} = (0.008 - 0.0039) + 0.002 = 0.0061$$

(You can also use the expression based on concentration in the recycle-reactor. Be careful with the nomenclature used for C_{Ao} vs C_{Af} and C_{AP} used here and in the lecture, and also that for v_o, v_o' and v_r, etc.)

Example 2

A MFR and PFR of equal volumes are connected in parallel for carrying a first-order ($k = 1 \text{ min}^{-1}$) liquid phase reaction. The concentration of the reactant in feed is 1 M. In a normal SS operation, the valves A, B, and C are open but D is closed. The concentration of A in the product stream is measured to be 0.1 M. On some day, MFR stops malfunctioning. The plant operator immediately shuts off the valves A & B, and diverts the entire feed ($Q = 10 \text{ lpm}$) to the PFR by partly opening the valve D. Note that this operation will not work the same as before, because the exit concentration will be greater than 0.1 M under SS condition. In such case, you should, however, calculate the flow rate of the recycled stream which the operator should use so that the reactant concentration in the product stream does not exceed 0.2 M.

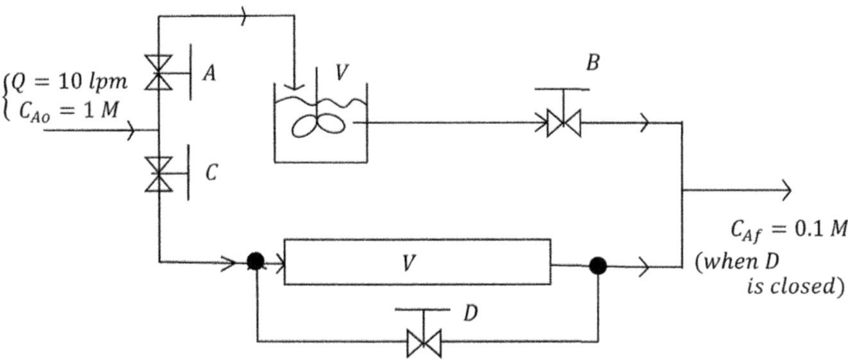

Ans:

Case 1:   When valve D is closed; no recycle

$$\tau_m = \frac{V}{Q_m} = \frac{C_{Ao} - C_{Af}}{(-r_A)_f} = \frac{1.0 - 0.1}{1.0 \times 0.1} = 9 \text{ min } (-r_A = kC_A)$$

$$\tau_p = \frac{V}{Q_P} = -\int_{C_{Ao}}^{C_{Af}} \frac{dC_A}{(-r_A)} = -\int_{1.0}^{0.1} \frac{dC_A}{C_A} = -\ln\left(\frac{0.1}{1.0}\right) = 2.3 \text{ min}$$

$$Q = Q_m + Q_P = 10 \Longrightarrow \frac{V}{9} + \frac{V}{2.3} = 10 \Rightarrow V = 18.31\ L$$

Case 2:   Recycle; D is open, B is closed, $C_{Af} = 0.2\ M$, we have tubular reactor with recycle.

$$\tau_R = \frac{V}{Q} = -(R+1)\int_{C_{A1}}^{C_{Af}} \frac{dC_A}{C_A}$$

C_{A1} is the concentration of mixed stream at recycle-feed to the reactor. Check the lecture.

$$C_{A1} = \frac{F_{Ao} + RF_{Af}}{v_o + Rv_f} = \frac{C_{Ao}v_o + RC_{Af}v_o}{v_o + Rv_o} \quad (v_f = v_o),\ \varepsilon_A = 0$$

$$= \frac{C_{Ao} + RC_{Af}}{1 + R} = \frac{1 + R \times 0.2}{1 + R}$$

Therefore,

$$\frac{18.31}{10} = -(R+1) \times \ln \frac{C_{Af}}{C_{A1}}$$

$$= (R+1)\ln\left(\left(\frac{0.2R+1}{R+1}\right) \times \frac{1}{0.2}\right)$$

$$1.83 = (R+1)\ln\left(\frac{R+5}{R+1}\right)$$

$$R =\sim 0.3 \text{ (requires iteration)}$$

$$v_{recycled} = R \times v_f = 0.3 \times v_o = 0.3 \times 10 = 3 \text{ lpm}$$

Autocatalytic Reactions (Or, Varying Order-Reactions?)

11

$$A + R \longrightarrow R + R \quad \text{or} \quad A \xrightarrow{+R} R; \ (-r_A) = kC_A C_R$$

- Microbial or enzymatic reactions.
- Chemical reactions are uncommon.
- Serves as a good example of non-elementary reactions having varying rate-order.

$A + R \rightarrow R + R$ (recycling is must)

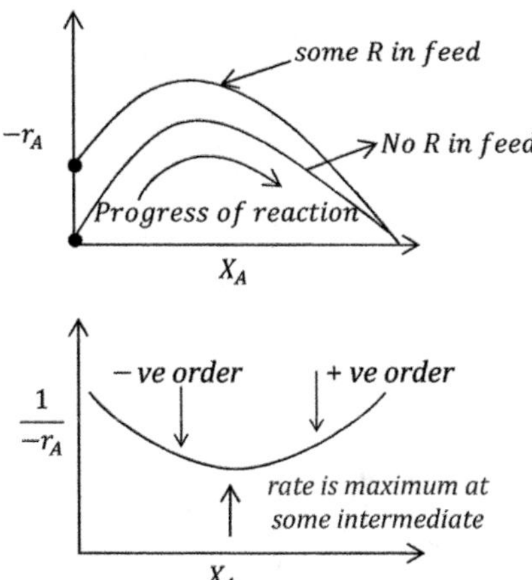

© The Author(s) 2025

N. Verma, *Chemical Reaction Engineering*,

https://doi.org/10.1007/978-3-031-88691-1_11

PFR Versus CSTR:

(a) X_d (desired/outlet) $\leq X_{min}$

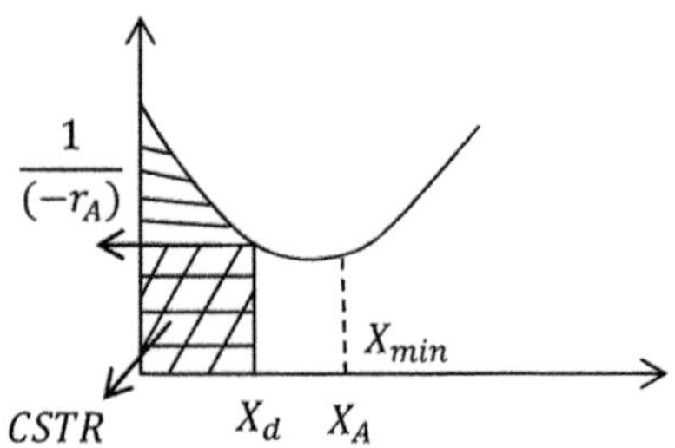

(b) $X_d \gg X_{min}$ (Depending on the kinetic rate, $V_{PFR} > V_{CSTR}$ or $V_{CSTR} > V_{PFR}$)

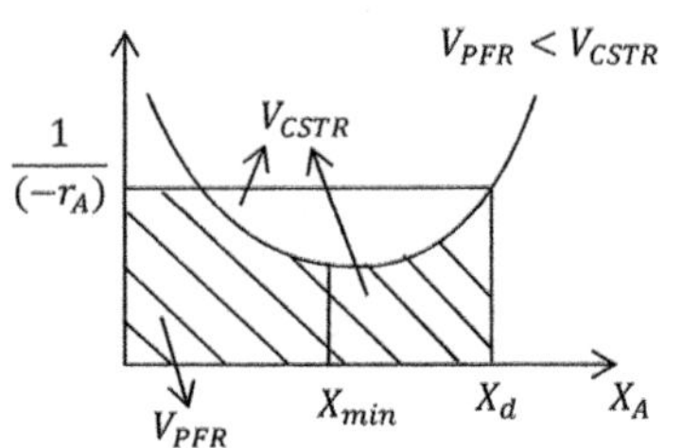

(c) CSTR followed by PFR (this is the best)

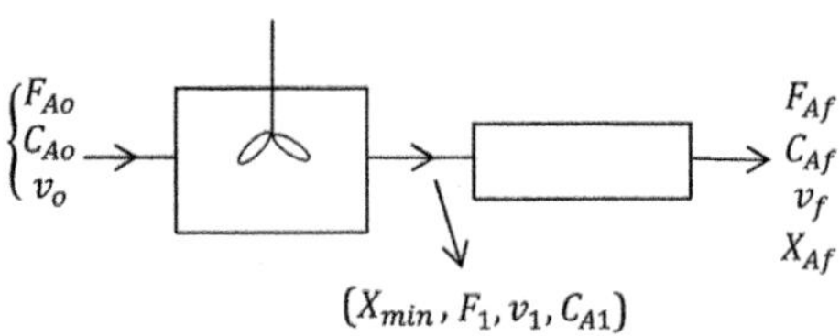

$$(X_{min}, F_1, v_1, C_{A1})$$

$$V_{min} = V_{CSTR} + V_{PFR}$$

(d) Reaction followed by separation (even better, cost of separator permitting)

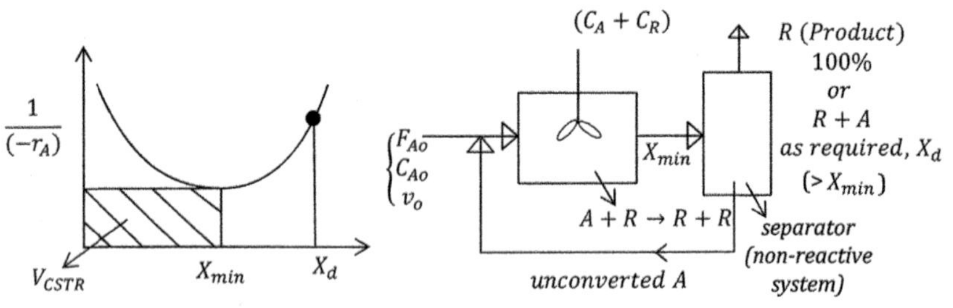

Example: We wish to explore various reactor setups for the transformation of A into R. The feed contains 99% A, 1 % R; the elements desired product is to consist of 10% A, 90 % R.

$$(A + R \rightarrow R + R); \quad k = 1\frac{\text{lit}}{\text{mol-min}}$$

$$\left(C_{Ao} + C_{Ro} = C_A + C_R = C_A + C_R = C_o = \frac{1 \text{ mol}}{\text{lit}}\right) \quad \text{(given)}$$

$$C_R = 0.9\frac{\text{mol}}{\text{lit}}$$

(a) Plug flow;
(b) Mixed flow;
(c) Minimum size set-up without recycle and
(d) Reaction followed by separation.

Ans:

$$- r_A = kC_A C_R = kC_{Ao}(1 - X_A)C_{Ao}X_A$$
$$= kC_{Ao}^2 X_A(1 - X_A)$$

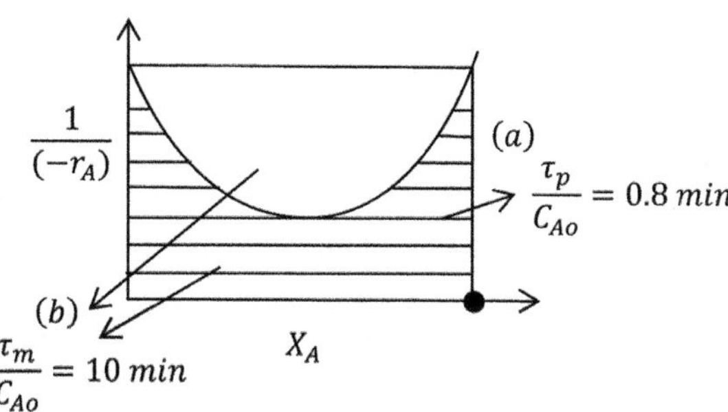

$$C_{Rf} = 0.9 \Rightarrow C_{Af} = 0.1$$
$$C_{Af} = C_{Ao}(1 - X_A)$$
$$0.1 = 0.99(1 - X_A)$$
$$X_A = 0.89$$

(c)

$$\tau_{\min(m)} + \tau_p = 2 + 2.2 = 4.2\ min \quad \text{(see figure next)}$$

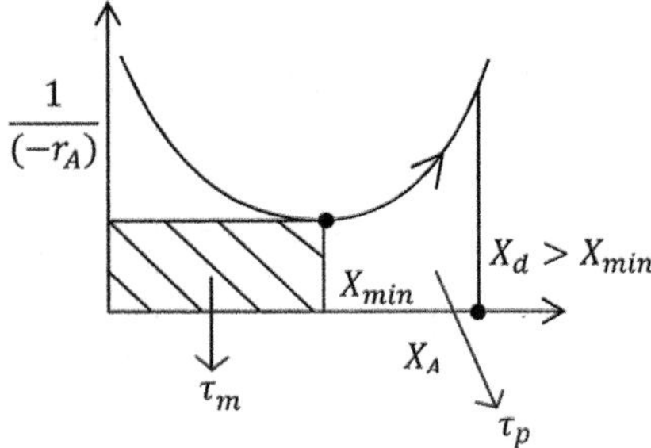

(d) Reaction followed by separation.

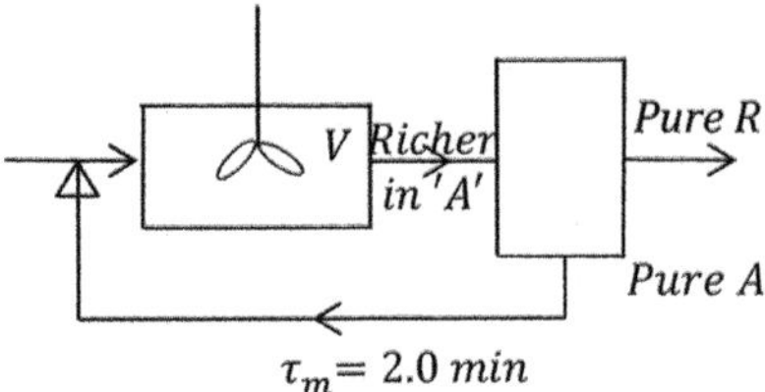

Recycling of an autocatalytic (or varying-order) reaction

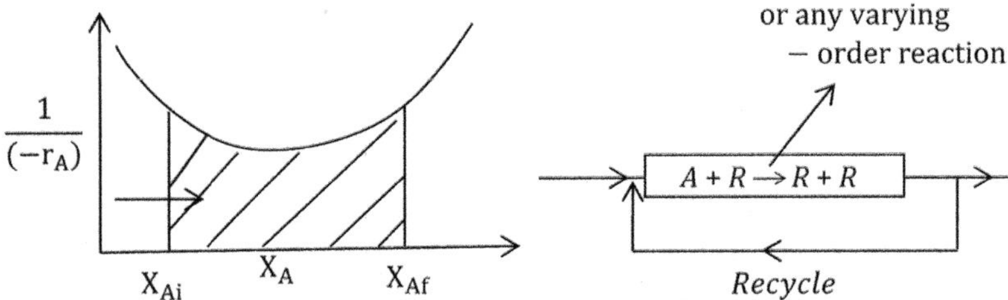

From the discussion of the preceding chapters, it is clear that the recycling of a tubular reactor (PFR) creates back-mixing, and increasing recycle ratio will, therefore, increase the back-mixing, lowering the average concentration in the reactor. Thus, it is also clear that, for a +ve order reaction the recycling or increasing recycle ratio will hurt the reactor performance, as 'PFR' gradually moves towards 'MFR'. On the other hand recycling will help (improve) a −ve order reaction. In either case (+ve or −ve order reaction), recycling will have a continuously increasing or decreasing effect on the performance of the reactor. In such case, no optimization is required. How about the reaction like the above shown in the figure? +ve order followed by −ve order, or vice-versa?

One has to optimize 'R' the recycle ratio

Revisit the model equation for recycling:

$$\frac{\tau}{C_{Ao}} \equiv \int\limits_{X_{Ai}=\frac{RX_{Af}}{R+1}}^{X_{Af}} \frac{R+1}{(-r_A)}\, dX_A$$

'R' effects 'X_{Ai}'. Therefore, set

$$\frac{d\,(\tau/C_{Ao})}{dR} = 0$$

to determine 'R' or the corresponding 'C_{Ai}', the mixing streams concentration. Note that it is an uncommon integral for differentiation, as the lower limit also contains 'R'. Recall Leibnitz rule or theorem of such unusual differentiation. If

$$F\,(R) = \int\limits_{a(R)}^{b(R)} f\,(X,\,R)\,dX$$

Then

$$\frac{dF}{dR} = \int\limits_{a(R)}^{b(R)} \frac{\partial f\,(X,\,R)}{\partial R}\,dX + f\,(b,\,R)\,\frac{db}{dR} - f\,(a,\,R)\,\frac{da}{dR}$$

In the present case, we have a simple function with simple limits!

The answer is:

$$\frac{1}{(-r_A)}\bigg|_{X_{Ai}} = \frac{\int_{X_{Ai}}^{X_{Af}} \frac{dX_A}{(-r_A)}}{(X_{Af} - X_{Ai})}$$

But, it requires an iterative technique: At what X_{Ai} (the mixing conversion) LHS = RHS for given X_{Af}? See the book. Graphical technique will work by determining various areas under the curve, $\frac{1}{(-r_A)}$ versus X_A.

The best approach is to write a simple code (on Matlab) to determine 'X_{Ai}' or R by mixing rule:

$$v_o C_{Ao} + v_R C_{Af} = (R+1)\,v_o C_{Ai} \quad (\varepsilon_A = 0)$$

The latter exercise is the same as in the previous chapter.

Parallel Reactions

Size determination, product distribution

$$A \xrightarrow[k_2]{k_1} \begin{array}{l} B \ (\textit{desired}) \\ C \ (\textit{unwanted side reaction}) \end{array}$$

Product distribution

$$A \xrightarrow{k_1} R; \quad r_R = k_1 C_A^{a_1} : \text{desired}$$

$$A \xrightarrow{k_2} S; \quad r_S = k_2 C_A^{a_2} : \text{undesired}$$

$$(-r_a) = \left(k_1 C_A^{a_1} + k_2 C_A^{a_2}\right)$$

$$\text{Ratio} = \frac{r_s}{r_R} = \left(\frac{k_2}{k_1}\right) C_A^{(a_2-a_1)} = \left(\frac{dC_s}{dC_R}\right) \text{ or } \frac{\Delta C_s}{\Delta C_R}$$

Ratio to be minimum to keep the unwanted species at the minimum level.

(a) If $a_2 > a_1$: 'C_A' is kept minimum by using CSTR.
(b) If $E_2 > E_1$: reactor temperature should be looked at.

© The Author(s) 2025

N. Verma, *Chemical Reaction Engineering*,

https://doi.org/10.1007/978-3-031-88691-1_12

Recall 1st/2nd lecture:

Must choose lower temperature;

$$\frac{E_2}{RT} > \frac{E_1}{RT}, \text{ as}$$

High T favors the reaction having larger E.

Low T favors the reaction having smaller E.

Similarly,

$$A + B \xrightarrow{k_1} R, \text{ desired}; \quad r_R = \frac{dC_R}{dt} = k_1 C_A^{a_1} C_B^{b_1}$$

$$A + B \xrightarrow{k_2} S, \text{ unwanted}; \quad r_s = \frac{dC_s}{dt} = k_2 C_A^{a_2} C_B^{b_2}$$

$$\text{Ratio} = \frac{r_s}{r_R} = \frac{k_2 C_A^{a_2} C_B^{b_2}}{k_1 C_A^{a_1} C_B^{b_1}} = \left(\frac{k_2}{k_1}\right) C_A^{(a_2-a_1)} C_B^{(b_2-b_1)}$$

Different cases:

If $\quad a_2 > a_1 \quad b_2 > b_1; \quad C_A \downarrow \quad C_B \downarrow$: use CSTR

$$\left.\begin{array}{l} a_2 > a_1 \quad b_1 > b_2; \quad C_A \downarrow \quad C_B \uparrow \\[2mm] a_1 > a_2 \quad b_2 > b_1; \quad C_A \uparrow \quad C_B \downarrow \end{array}\right\} : \text{use PFR with side-entry}$$

$\quad a_1 > a_2 \quad b_1 > b_2; \quad C_A \uparrow \quad C_B \uparrow$: use PFR

Case (1):

Case (2/3):

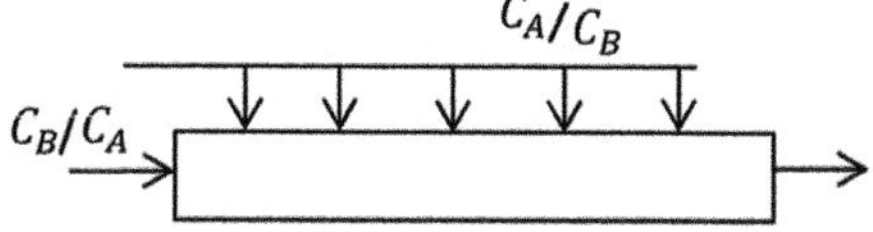

Case (4):

"A high reactant concentration favors the reaction of higher order, a low concentration favors the reaction of lower order".

Quantitative treatment of product distribution and reactor size

(a) 'Instantaneous' fractional yield of R,

$$\psi = \left(\frac{\text{moles of } R \text{ formed}}{\text{moles of } A \text{ reacted}} \right) = \left(\frac{dC_R}{-dC_A} \right)$$

$$= \left(\frac{dC_R}{-dC_A} \right)_{PFR} \quad \text{or} \quad \left(\frac{\Delta C_R}{-\Delta C_A} \right)_{CSTR}$$

where, ψ is positive.

(b) $\phi = $ overall fractional yield of R

$$+ve = \frac{C_{Rf}}{C_{Ao} - C_{Af}} = \frac{C_{Rf}}{(-\Delta C_A)}$$

PFR:

$$\phi_p = -\frac{1}{C_{Ao} - C_{Af}} \int_{C_{Ao}}^{C_{Af}} \underbrace{\psi \cdot dC_A}_{dC_R} = \frac{1}{\Delta C_A} \int_{C_{Ao}}^{C_{Af}} \psi \cdot dC_A = \left(\frac{\text{all } R \text{ formed}}{\text{all } A \text{ reacted}} \right)$$

MFR:

$$\phi_m = \psi_{exit} = \frac{C_{Rf}}{C_{Ao} - C_{Af}}$$

MFRs in series:

$$C_{Ao} \longrightarrow \boxed{\psi_1} \longrightarrow \boxed{\psi_2} \longrightarrow \boxed{\psi_3} \longrightarrow C_{Af}$$

$$\phi_m = \frac{C_{Rf}}{C_{Ao} - C_{Af}}$$

$$\psi_1 = \frac{C_{R1}}{C_{Ao} - C_{A1}},$$

$$\psi_2 = \frac{C_{R2} - C_{R1}}{C_{A1} - C_{A2}},$$

$$\vdots$$

$$\psi_N = \frac{C_{RN} - C_{RN-1}}{C_{AN-1} - C_{AN}}$$

$$\psi_1 \left(C_{Ao} - C_{A1}\right) + \psi_2 \left(C_{A1} - C_{A2}\right) + \cdots + \psi_N \left(C_{AN-1} - C_{AN}\right)$$

$$= C_{RN} = \phi_{N,mix}\left(C_{Ao} - C_{Af}\right)$$

Note that ψ or ϕ is a convenient (auxillary) parameter for studying parallel reactions. In other words, ψ or C_A plot is used for determining product distribution, similar to $1/(-r_A)$ versus C_A used for sizing the reactor.

(1) PRF:

(2) CSTR:

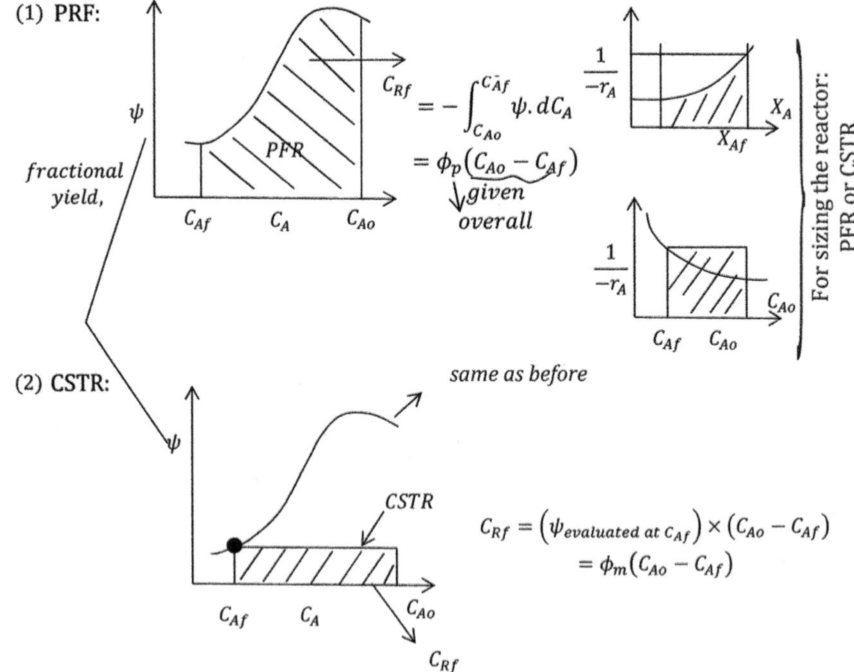

(3) CSTR in series: (4) CSTR followed by PFR:

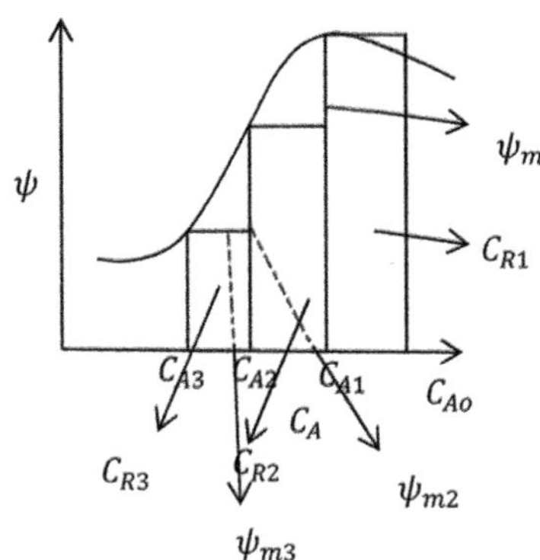

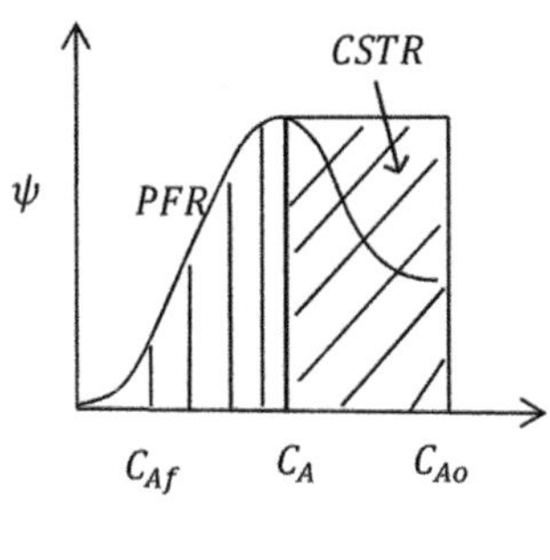

Example 1

Under ultraviolet radiation, a reactant ($C_{Ao} = 10.0$ kmoles/m^3) in a process stream ($v = 1$ m^3/min) decomposes as follows:

$$r_R = 16\,C_A^{0.5} \quad \frac{kmol}{m^3-min}$$

$$A \Big\langle \begin{matrix} R \\ S \\ T \end{matrix}$$

$$r_S = 12\,C_A \quad \frac{kmol}{m^3-min}$$

$$r_T = C_A^2 \quad \frac{kmol}{m^3-min}$$

R is the desired material. We wish to design a reactor set up (either MFR or PFR) for the process. If the concentration of A at the reactor-exit is to be 0.2 kmoles/m^3, calculate the concentration of R in the product stream, as well as the volume of the reactor which you will select. (Levenspiel-book)

Ans:

$$\psi\left(\frac{R}{A}\right) = \frac{16C_A^{0.5}}{16C_A^{0.5} + 12C_A + C_A^2}$$

$$= \frac{16}{16 + 12C_A^{0.5} + C_A^{1.5}}$$

$$= \left(\frac{dC_R}{-dC_A}\right)_{PFR}$$

$$= \left(\frac{\Delta C_R}{-\Delta C_A}\right)_{MFR}$$

Plot

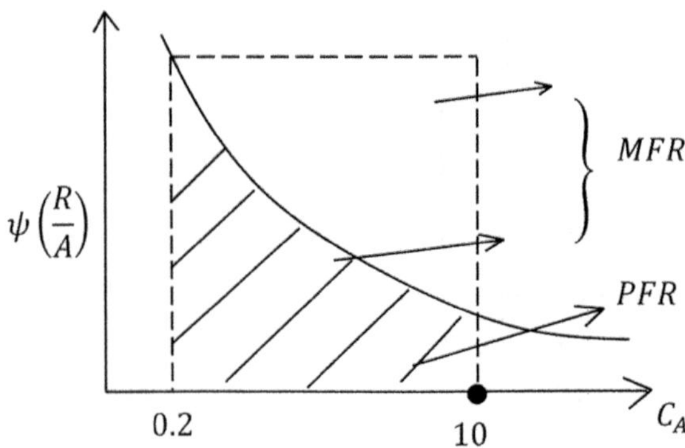

From the graph (ψ decreases with increasing concentrations) above, it is clear that MFR will deliver higher product (R) distribution or yield than PFR. See the area covered under the curve for two reactors.

MFR:

$$\Delta C_R = (-\Delta C_A)\,\psi_{exit}$$

or

$$(-\Delta C_A)\,\psi_f$$

$$C_{Rf} - C_{Ro} = (C_{Ao} - C_{Af})\left(\frac{16}{16 + 12C_{Af}^{0.5} + C_{Af}^{1.5}}\right)$$

Given:

$$C_{Ao} = 10,\ \ C_{Af} = 0.2,\ \ C_{Ro} = 0 \implies C_{Rf} = 7.308 \text{ kmoles/m}^3$$

(Note: $\psi_{exit} = 0.7457$).

Volume of the reactor will still come from the design equation (do not forget)

$$\tau_{MFR} = \frac{V}{v_o} = \frac{-\Delta C_A}{(-r_A)_f}$$

$$V = \frac{1 \times (10 - 0.2)}{(16 \times 0.2^{0.5} + 12 \times 0.2 + 0.2^2)}$$
$$= 1.0213 \text{ m}^3$$

$$\left(-r_A = -\frac{dC_A}{dt} = 16\,C_A^{0.5} + 12C_A + C_A^2\right)$$

Example 2

A liquid-phase chemical reaction $A \rightarrow R$ is carried out in a MFR under SS condition. A side product S is also formed in the parallel reaction $A \rightarrow S$. The following data are measured under different conditions:

τ (min)	10	30	8	1	4	4	20
C_{Ao} (moles/m^3)	2	6	6	5	24	11	16
C_A (moles/m^3)	0.5	1	2	3	4	6	8
C_S (moles/m^3)	0.33	1.25	0.89	0.38	3.2	0.61	0.79

The same reaction as above is now carried out in a MFR followed by PFR. The volumetric flow rate of feed is 1 m^3/min. The concentration of A in the feed is 6 moles/m^3. The required concentration of A at the exit of the PFR is 0.5 moles/m^3.

(a) Determine the minimum total reactor-volume for this configuration (MFR-PFR)
(b) Determine the concentration of S at the exit of the reactor in the configuration of part (a).
(c) If you are allowed to select reactors of different volumes, determine the total volume of this configuration (MFR followed by PFR) to maximize the concentration of S at the exit of PFR:

Ans: Plot $\left(\dfrac{1}{-r_A}\right)$ versus C_A and ϕ versus C_A. Create a table:

$$\frac{1}{(-r_A)_f} = \frac{\tau}{(C_{Ao} - C_{Af})} \Rightarrow \text{Fill the table}$$

τ	C_{Ao}	C_A	C_S	$1/(-r_A)$	ϕ
10	2	0.5	0.33	6.67	0.222
30	6	1	1.25	6.0	0.250
8	6	2	0.89	2.0	0.222
1	5	3	0.38	0.5	0.187
4	24	4	3.2	0.2	0.160
4	11	6	0.61	0.8	0.122
20	16	8	0.79	2.5	0.099

$$\tau_{MFR} = \frac{V}{v_o} = -\frac{\Delta C_A}{(-r_A)_f}$$

$$= \frac{C_{Ao} - C_{Af}}{(-r_A)_f}$$

$$\phi_{\frac{S}{A}} = \left(\frac{C_S}{C_{Ao}-C_A}\right)_f \quad as \quad A{\Large\diagup\atop\diagdown}{\!\!\raisebox{0.4ex}{R}\atop\raisebox{-0.4ex}{S}}$$

$\Rightarrow$ Fill up the table.

(a) Reactor volume comes from $\frac{1}{(-r_A)}$ versus C_A (Recall)

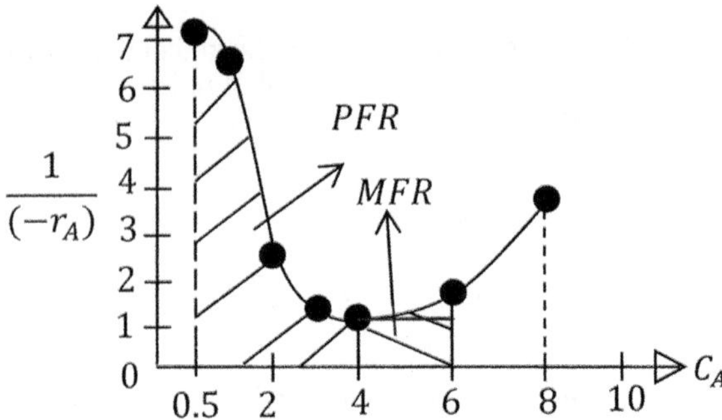

Area under the curve from $C_{Ao} = 6$ to $C_{Af} = 0.5$ will give

$$\tau = \tau_{MFR} + \tau_{PFR}$$

$$C_A: \quad (6-4) \quad (4-0.5)$$

$$\tau_{MFR} = \frac{-\Delta C_A}{(-r_A)_f}$$
$$= (6-4) \times 0.2$$
$$= 0.4 \text{ min}$$

τ_{PFR} should be calculated using the Trapezoidal rule of integration:

$$\tau_{PFR} = -\int_4^{0.5} \frac{dC_A}{(-r_A)}$$
$$= \sum_{1,2,3,4} \frac{\Delta C_A}{(-r_A)\,avg}$$
$$= (4-3)\left(\frac{0.2+0.5}{2}\right) + (3-2)\left(\frac{0.5+2}{2}\right)$$
$$+ (2-1)\left(\frac{2+6}{2}\right) + (1-0.5)\left(\frac{6+6.66}{2}\right)$$
$$= 8.765$$

$$\tau = 0.4 + 8.765 = 9.165 \text{ min}$$

$$V = 9.165 \times 1 = \underline{9.165} \text{ m}^3$$

(b) Plot ψ versus C_A

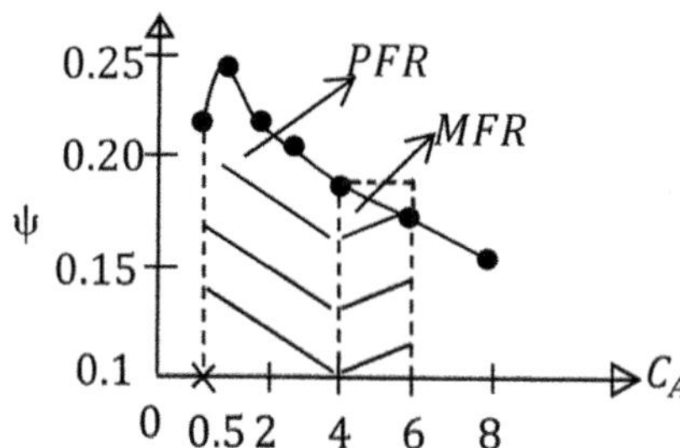

$$\left. \begin{array}{l} C_s = (\psi)_f \times (6-4) + \int\limits_{4}^{0.5} \phi(-dC_A) \\[2mm] = 0.16 \times 2 + \sum_{1,2,3,4} \phi_{avg}(\Delta C_A) \end{array} \right\} \text{area under curve}$$

$$C_s = 0.32 + \frac{(0.16 + 0.1875)}{2} \times 1 + \frac{(0.1875 + 0.222)}{2} \times 1$$
$$+ \frac{(0.222 + 0.25)}{2} \times 1 + \frac{(0.25 + 0.222)}{2} \times 0.5$$

Note, again the evaluation of the integral $\int\limits_{4}^{0.5} \phi(-dC_A)$ uses Trapezoidal rule (numerical), i.e.,

$$\Delta X \times \left(\frac{f(a) + f(b)}{2} \right)$$

where ΔX is the step size

$$C_s = 0.32 + 0.7325 = 1.052 \text{ moles/m}^3$$

(c) Maximum 'C_s' from the configuration of a MFR followed by PFR will be the highest area from ϕ versus C_A curve. That is, MFR will operate from $C_{Ao} = 6.0$ to $C_{Af} = 1.0 \text{ moles/m}^3$; ϕ being the maximum (0.25 moles/m^3) at 1.0 moles/m^3, then PFR from $C_A = 1.0$ to 0.5 moles/m^3.
Thus,

$$\tau_{CSTR} = \frac{-\Delta C_A}{(-r_A)_f} = (6 - 1.0) \times 6 = 30 \text{ min}$$

$$\tau_{PFR} = -\int\limits_{1.0}^{0.5} \frac{dC_A}{(-r_A)} = (1 - 0.5) \times \left(\frac{6 + 6.667}{2} \right)$$
$$= 3.167 \text{ min}$$

$$V_{total} = (30 + 3.167) \times 1 = 33.16 \text{ m}^3$$

(But, note that this is not the V_{min}. You cannot have both $C_{s,max}$ (max·yield) and minimum volume of the reactors!)

In this case, however,

$$C_s \text{ is maximum} = (6 - 1) \times 0.25 + (1 - 0.5) \times \left(\frac{0.25 + 0.222}{2}\right)$$

$$\downarrow$$

$$(\text{here, } \phi_{MFR} \text{ at } 1 \text{ mole/m}^3)$$

$$= 1.25 + 0.118$$

$$= \underline{1.368} \text{ moles/m}^3$$

Example 3: (Levenspiel)

Consider the aqueous reactions

$$A + B \quad \begin{cases} \xrightarrow{k_1} R(desired) & \dfrac{dC_R}{dt} = C_A^{1.5} C_B^{0.3} \\[2mm] \xrightarrow{k_2} S \ (undesired) & \dfrac{dC_S}{dt} = C_A^{0.5} C_B^{1.8} \end{cases} mole/l\text{-}min$$

For 90% conversion of A, find the concentration of R in the product stream. Equal volumes of A and of B streams are fed to the reactor, and each stream has concentration of 20 moles/l of reactant.

The flow in the reactor follows: (a) Plug flow (b) Mixed flow (c) Plug flow A - side entry (mixed flow) B.

Ans:

$$\psi\left(\frac{R}{A}\right) = \frac{dC_R}{\underbrace{dC_R + dC_S}_{-dC_A}} = \frac{C_A^{1.5} C_B^{0.3}}{C_A^{1.5} C_B^{0.3} + C_A^{0.5} C_B^{1.8}} = \frac{C_A}{C_A + C_B^{1.5}}$$

(a) PFR:

$$C_{Ao} = C_{Bo} = 10 \frac{mol}{l}$$

(equal volumetric flow-streams are mixed; dilution effect).

$$C'_{Ao}$$
$$C'_{Bo}$$

$$C_{Ao} = C_{Bo} = 10 \ mol/lt \qquad\qquad C_{Af} = C_{Bf}$$
$$C_{Rf}, C_{Sf}$$

Stoichiometry suggests $C_A = C_B$ everywhere ($C_{Ao} = C_{Bo}$)

$$C_{Af} = 1.0 \text{ moles/l } (90\% \text{ conversion}) = C_{Bf}$$

$$\phi_p = -\frac{1}{C_{Ao} - C_{Af}} \int \psi\, dC_A$$

$$= -\frac{1}{10-1} \int_{10}^{1} \left(\frac{C_A}{C_A + C_A^{1.5}}\right) dC_A$$

$$= 0.164 \quad (\text{check the book})$$

$$C_{Rf} = 0.164 \times (10 - 1) = 1.476 \text{ mol/l}$$

One can define

$$\psi\left(\frac{S}{A}\right) \implies \frac{dC_s}{dC_R + dC_S} \implies \phi_p$$

which will be $(1 - 0.164) = 0.836$

$$C_s = 0.836 \times (10 - 1) = 7.524 \text{ mol/l}$$

(b) Mixed flow

$$C_{Ao'} = C'_{Bo}$$
$$= 20 \; mol/lt$$

$$C_{Ao} = C_{Bo}$$
$$= 10 \; mol/l$$

$$C_{Af} = 1 \; mole/l$$
$$= C_{Bf}$$

$$\phi_m\left(\frac{R}{A}\right) = \psi_m\left(\frac{R}{A}\right)_{exit}$$

$$= \left(\frac{1}{1 + C_A^{0.5}}\right)_{exit}$$

$$= 0.5$$

$$C_{Rf} = 0.5 \times \left(C_{Ao} - C_{Af}\right) = 0.5\,(10 - 1)$$
$$= 4.5 \text{ mol/l}$$

$$C_{Sf} = (1 - 0.5) \times (10 - 1)$$
$$= 4.5 \text{ mol/l}$$

(c) Plug flow A – Mixed flow B

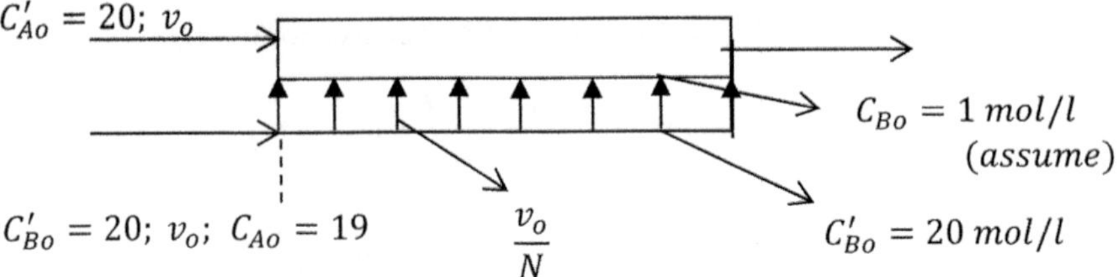

(1) Note assume that the side entry streams are sufficiently large in number so that the concentration of 'B' at the entry to the PFR is diluted to 1 mole/l everywhere.

(2) Considering that we have a PFR, the radial concentration gradient is negligible, and $C_B = 1$ mole/l throughout PFR.

Thus, the batch kinetics with $C_B \sim 1$ mol/l:

$$\frac{dC_R}{dt} = C_A^{1.5}; \quad \frac{dC_S}{dt} = C_A^{0.5}; \quad \psi = \left(\frac{R}{A}\right) = \frac{C_A}{1 + C_A}$$

$$\phi\left(\frac{R}{A}\right) = -\frac{1}{C_{Ao} - C_{Af}} \int_{C_{Ao}}^{C_{Af}} \psi_{\frac{R}{A}}\, dC_A = -\frac{1}{(19 - 1)} \int_{19}^{1} \frac{C_A\, dC_A}{(1 + C_A)}$$

(*Note*: 'C'_{Ao} also decreases to 19 mole/l from 20 mole/l on mixing with 'B' at the entrance.)

In other words,

$$C_{Rf} + C_{Sf} = 9$$

always holds good for 90% conversion of A.

$$\phi\left(\frac{R}{A}\right) = 0.87 \quad \text{(check the book for integration)}$$

Therefore,

$$C_{Rf} = \left(C_{Ao} - C_{Af}\right)\phi = (10 - 1) \times 0.87 = 7.85\ \text{mole/l}$$
$$C_{Sf} = (10 - 1) \times (1 - 0.87) = 1.15\ \text{mole/l}$$

Carefully note that in the third case as well, $C_{Ao} = 10$ and $C_{Af} = 1$ mole/l remain the same as before! Why? Note the increment in the volumetric flow rate in PFR because of the side entry of 'B' along the PFR length.

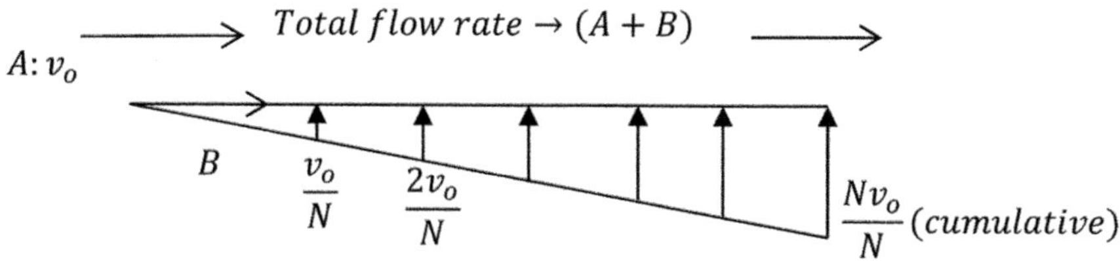

Here also, there is the 'dilution' effect as in the previous two cases, upon the mixing of 'A' and 'B' at the entrance to the reactor (PFR or MFR). The net effect is

$$\Delta C_A = (10 - 1) \ \text{mole/l}$$

The volume (l) used in the denominator of 'C_A' itself changes along the length

$$\Longrightarrow \Delta C_A = (10 - 1) \equiv (19 - 10) \ \text{mole/l}$$

On the other hand, 'ψ' is always local to the volume. This is a good example on how you can mix one component in small quantity relative to the other component in flow reactor, and still maintaining a high yield of the desired product. The assumption of C_B being constant throughput the length may be a bit difficult to justify or maintain in a real reactor.

Reactions in Series

$$A \xrightarrow{k_1} R \xrightarrow{k_2} S$$

How to maximize C_R if R is the desired product? PFR or CSTR?

(No need to define ψ or ϕ (instantaneous or total fractional yield))

PFR/Batch

$$\frac{dC_A}{dt} = r_A = -k_1 C_A$$

$$\frac{dC_R}{dt} = k_1 C_A - k_2 C_R$$

$$\frac{dC_S}{dt} = k_2 C_R$$

On integration,

$$C_A = C_{Ao} e^{-k_1 t}$$

$$C_R = \frac{k_1}{k_1 - k_2} \left(\exp\left(-k_2 t_1\right) - \exp\left(-k_1 t_1\right) \right)$$

$$C_S = C_{Ao} - C_A - C_R \quad \text{(from stiochmetry, or}$$

$$\frac{d}{dt}(C_A + C_R + C_S) = 0 \Longrightarrow C_A + C_R + C_S = C_{Ao})$$

© The Author(s) 2025

N. Verma, *Chemical Reaction Engineering*,

https://doi.org/10.1007/978-3-031-88691-1_13

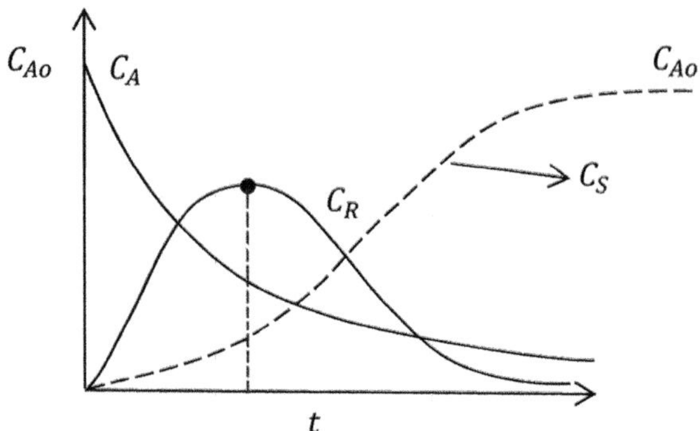

(1) 'C_R' goes through 'maximum concentration'.
(2) Two extreme situations:

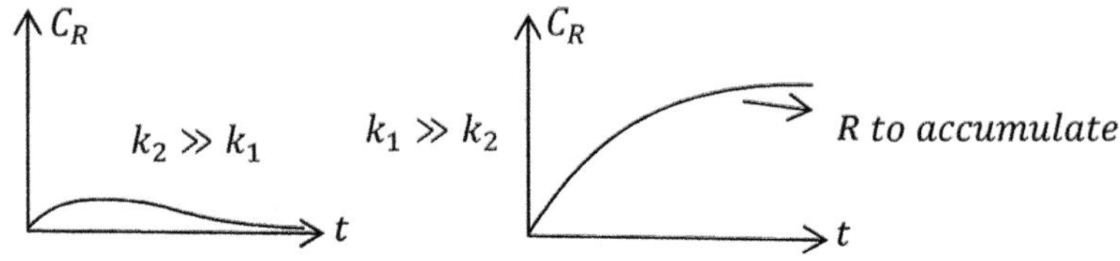

(R behaves like short-lived species)

One which is slower controls the overall reaction.

Check:

$$C_S = C_{Ao}\left(1 - e^{-k_1 t}\right) \quad \text{if } k_1 \ll k_2$$

$$= C_{Ao}\left(1 - e^{-k_2 t}\right) \quad \text{if } k_2 \ll k_1$$

(3) Peak in C_R occurs at

$$t_{\max} = \frac{1}{k_{log-mean}} = \left(\frac{\log\left(\frac{k_2}{k_1}\right)}{k_2 - k_1}\right) \quad \text{from} \quad \frac{dC_R}{dt} = 0$$

(4) Batch data:

C_R	t
⋮	⋮
$C_{R,max}$	t_{max}
⋮	⋮

$$C_{R,max} = C_{Ao} \left(\frac{k_1}{k_2}\right) @ \left(\frac{\log(k_1/k_2)}{k_1 - k_2}\right) \text{time}$$

Determine k_1 & k_2

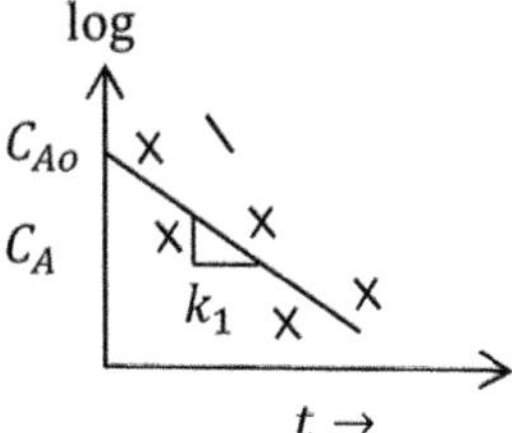

Once k_1 is determined from the batch data for C_A, determine k_2 from $C_{R,max}$ or t_{max}.

PFR:

$$C_{Ao}, C_{Ro} = C_{So} = 0$$

v_o
F_{Ao}
⋮
$'\tau'$
$\left.\begin{matrix} C_{Af} \\ C_{Rf} \\ C_{Sf} \end{matrix}\right\}$

$$X_A, F_A, F_R, F_S$$

Perform experiments for different values of v_0 or τ. Alternatively, take samples from the reactor along the axial direction at fixed v_0:

$$\tau_1 = \frac{V_1}{v_o}, \quad \tau_2 = \frac{V_2}{v_o}, \text{etc.}$$

Performance Equation

't' replaced by 'τ'

$$\tau = \frac{V}{v_o} = C_{Ao}\frac{dX}{(-r_A)}$$

Data:

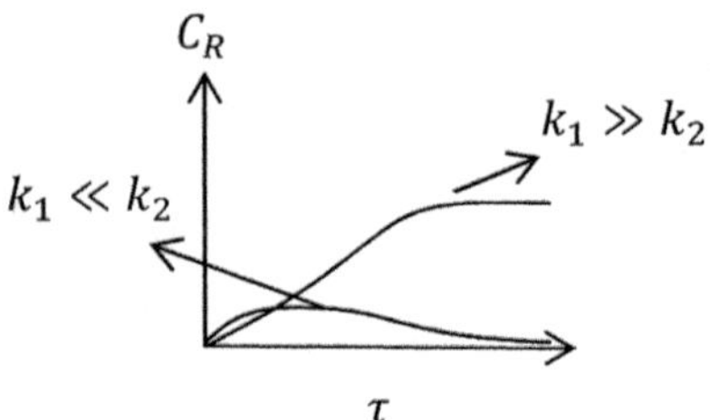

$$\left.\begin{array}{l}
C_{Af} = C_{Ao}\exp\left(-k_1\tau\right) \\[4pt]
C_{Rf} = \dfrac{k_1}{k_1-k_2}\left(\exp\left(-k_2\tau\right) - \exp\left(-k_1\tau\right)\right) \\[4pt]
C_{Sf} = C_{Ao} - C_{Af} - C_{Rf}
\end{array}\right\}$$

"Peak of C_R moves from the inlet to outlet as k_2/k_1 varies from a very large value to smaller value".

How do we have maximum C_{Rf} in a PFR for parallel-reaction?

(a) Analyse their 'E_S'; vary "temperature" to make rate vary with T. A reaction of higher 'E' will be favored by high temperature.

(b) Or choose a PFR of $\tau = \tau_{\max}$, so that

$$C_{Rf} = C_{Rf,\max} \implies \tau_{\max} = \frac{V}{v_o}$$

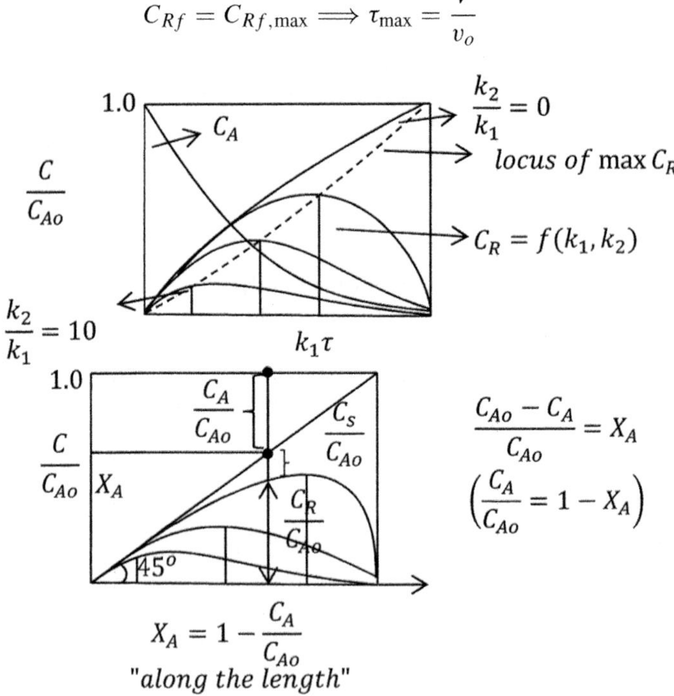

$$X_A = 1 - \frac{C_A}{C_{Ao}}$$

"along the length"

MFR/CSTR

$$\tau = \frac{V}{v_o} = C_{Ao}\frac{X_{Af}}{(-r_A)_f} = C_{Ao}\underbrace{\frac{X_{Af}}{kC_{Ao}(1-X_{Af})}}_{C_{Af}} \quad \longrightarrow \left(\frac{C_{Ao}-C_{Af}}{C_{Ao}}\right)$$

$$\frac{C_{Af}}{C_{Ao}} = \frac{1}{1+k_1\tau_m} \quad (\text{Inlet} - \text{Outlet} - \text{reaction} = 0)_{A,R,S}$$

$$\left.\begin{aligned}
\tau_m &= \frac{V}{v_o} = \frac{C_{Rf}}{(k_1 C_{Af} - k_2 C_{Rf})}\\
&= \frac{C_{Ao} - C_{Af}}{k_1 C_{Af}}\\[4pt]
\frac{C_{Rf}}{C_{Ao}} &= \frac{k_1\tau_m}{(1+k_1\tau_m)(1+k_2\tau_m)}\\
\frac{C_{Sf}}{C_{Ao}} &= \frac{k_1 k_2 \tau_m^2}{(1+k_1\tau_m)(1+k_2\tau_m)}\\
C_A &+ C_R + C_S = C_{Ao} = \text{Const}
\end{aligned}\right\} \text{Check}$$

Also,

can be shown

$$\tau_{m,\max} = \frac{1}{\sqrt{k_1 k_2}} \quad \left(\frac{dC_{Rf}}{d\tau_m}\right) = 0$$

Conclusions $A \to R \to S$ (irreversible, first order reaction)

1. For a given k_2/k_1 ratio, there is a $C_{R,\max}$ corresponding to '$\tau_{\max}$' either in PFR or CSTR.

2. $C_{R,\max}$, $\tau_{\max} = f(k_1, k_2)$; one can alter '$\tau$' to achieve desired $C_{R,\max}$.
3. The location of $C_{R,\max}$ can also alter by changing 'T' and making use of sensitivity of 'E' to change in temperature.
4. There is no need to use or work on 'ψ', although there is ψ or ϕ.
5. For a given reaction, $C_{R,\max}\big|_{\text{Plug-flow}} > C_{R,\max}\big|_{MFR}$ for the same τ.
6.

$$\text{Except for } \frac{k_1}{k_2} = 1, \ \tau_{\max}|_{PFR} < \tau_{\max}|_{CSTR}$$

7.

$$\phi_p \left(\frac{C_R}{C_A} \quad \text{or} \quad \frac{C_R}{C_R + C_S} \right) > \phi_m \left(\frac{C_R}{C_A} \quad \text{or} \quad \frac{C_A}{C_R + C_S} \right)$$

8. Qualitatively, it can be shown that back-mixing reduces the concentration of the intermediate species (R), which is consistent with the above finding in (5).

Note: back-mixing:—mixing of different fluids of different compositions.

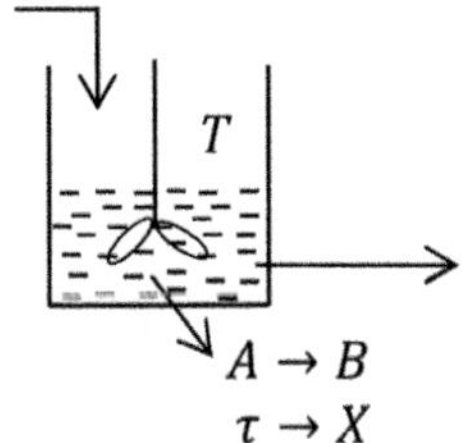

Heat of reaction: +ve or −ve or ~ 0

(a) Reversible and irreversible reaction (Kinetics)
(b) Exothermic and endothermic (Thermodynamics)
(c) Isothermal and non-isothermal (Operation)
(d) Steady state and non-steady state (Process)

Energy Barrier Diagram

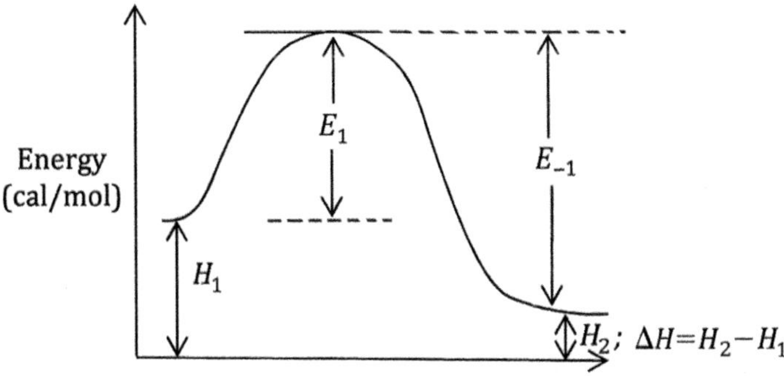

© The Author(s) 2025

N. Verma, *Chemical Reaction Engineering*,
https://doi.org/10.1007/978-3-031-88691-1_14

(1) No reaction is truly irreversible (it is an extreme case of a reversible reaction; when $r_f \gg r_b$).

(2) Thermodynamics decides if a reaction is exothermic or endothermic

  ΔH_r is $-$ve: exothermic

  ΔH_r is $+$ve: endothermic

  (it may also be negligible)

  If reversible, then there exists "X_e" (equilibrium) < 1.0

Revision

$$aA \rightarrow rR + sS$$

(1) ΔH_r at the standard temperature (usually $25°C$) is known or available or can be calculated, in principle, from thermochemical data on the heat of formation or heat of combustion.

(2)

$$\Delta H_{r_2} = \Delta H_{r_1} + \int_{T_1}^{T_2} \nabla C_p dT \left(\text{def}^n \text{ of } \Delta H_{r_2}\right)$$

where

$$\nabla C_p = rC_{pr} + sC_{ps} - aC_{pA}$$

$$\searrow$$

$$\frac{cal}{mol - K}$$

$$C_{p,A} = \alpha_A + \beta_A T + \gamma_A T^2, etc.$$

  (a) Heat of reaction at temperature T_1

  (b) Dependence of C_{ps} on T over $T_1 \sim T_2$

(3) If ΔH_r is $+ve$ at T_1, ΔH_r could be $-ve$ at T_2

(4) $X_e = f(T); \quad A \underset{k_2}{\overset{k_1}{\rightleftarrows}} B;$

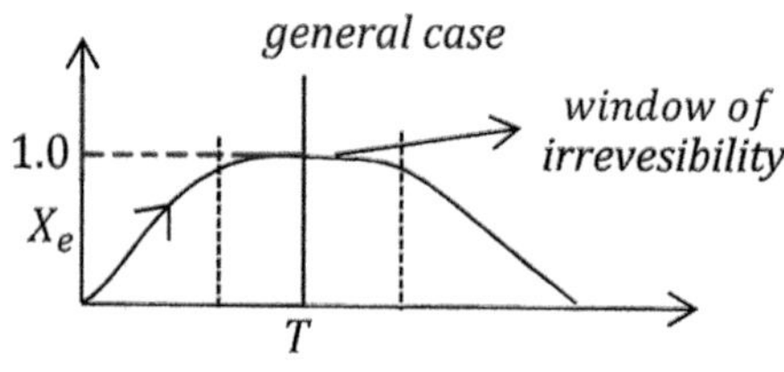

(5)

$$A \underset{k_2}{\overset{k_1}{\rightleftarrows}} B$$

$$K(T) = \frac{k_1(T)}{k_2(T)} = f(C_{Be}, C_{Ae})$$

Note:

$$k_1 \gg k_2; \quad K \gg 1 : \text{ irreversible}$$
$$k_1 \ll k_2; \quad K \ll 1 : \text{ no forward reaction}$$

(6)

$$rG_R^o + sG_S^o - aG_A^o = -RT \ln K = -RT \ln \frac{\left(\frac{f}{f_o}\right)_R^r}{\left(\frac{f}{f_o}\right)_A^o}$$

where G^o is the standard free energy.

Standard state:

Gases: (a) Pure, $P = 1$ atm

Solids: Pure unit pressure

Liquids: at its vapor pressure

Solution: $a = 1, 1M$

(7)

$$\frac{d(\ln K)}{dT} = \frac{\Delta H_r}{RT^2} \tag{14.1}$$

$$\ln \frac{K_2}{K_1} = -\frac{\Delta H_r}{R}\left[\frac{1}{T_2} - \frac{1}{T_1}\right]; \quad \Delta H_r = \text{ const over } T_1 \sim T_2$$

(8) Thermodynamic calculation:

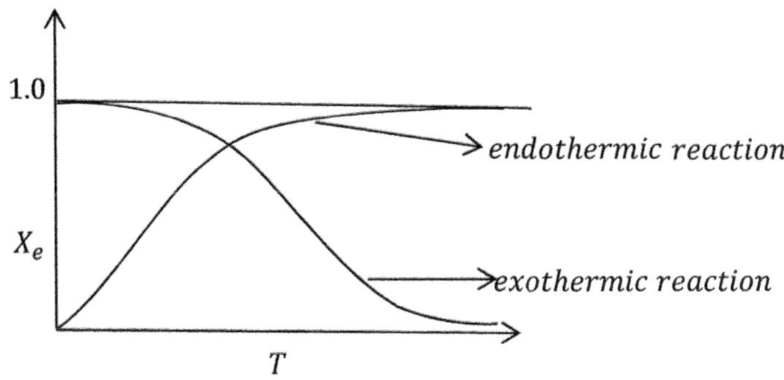

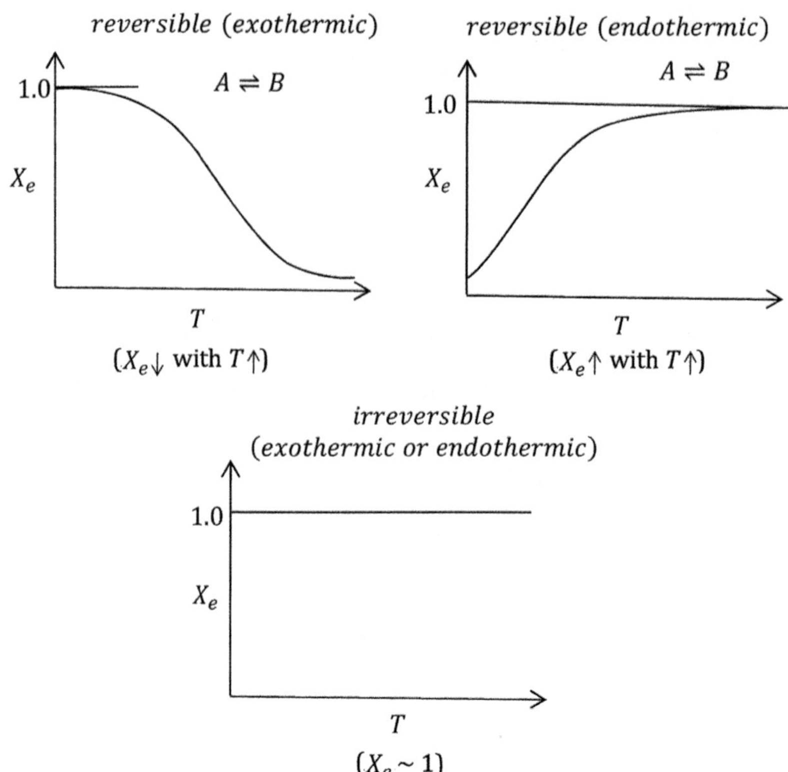

(9) Kinetics calculation: $A \rightleftharpoons B$

$$r = k_1 C_A - k_2 C_B$$

Example

Determine X_e $(0° - 100°C)$ for the elementary aqueous phase reaction.

$$A \underset{k_2}{\overset{k_1}{\rightleftharpoons}} R$$

(a) Given:

$$\Delta H^o_{r,298} = -75300 \text{ J/mol}$$

$$\Delta G^o_{298} = -14130 \text{ J/mol}$$

(b) C_p of A and $R = $ const.; $\Delta C_p = 0$ (assume)

(c)

$$\Delta H_{r,2} = \Delta H_{r,1} + \int_{T_1}^{T_2} \Delta C_p dT = \Delta H_{r,1}$$

$$\Delta H_r = \Delta H_{r,298} = -75300 \text{ J/mol}$$

$$K(T) = \frac{k_1}{k_2} = \frac{C_{Re}}{C_{Ae}} = \frac{X_e}{1 - X_e}$$

$$X_{Ae} = \frac{K(T)}{1 + K(T)}$$

(1) $\dfrac{d(\ln K)}{dT} = \dfrac{\Delta H_r}{RT^2}$

$$\ln \frac{K_2}{K_1} = \int_{T_1}^{T_2} \frac{\Delta H_r}{RT^2} dT = \frac{\Delta H_r}{R} \left[\frac{1}{T_1} - \frac{1}{T_2} \right]$$

(2) $\Delta G^o_f = -RT \ln K(T) \Rightarrow \ln K_{298} = \exp\left(-\frac{\Delta G^o}{RT}\right)$

St^d condition

$$= \exp\left(\frac{14,130}{8.314 \times 298}\right) = 300$$

Use (1) and (2) to plot:

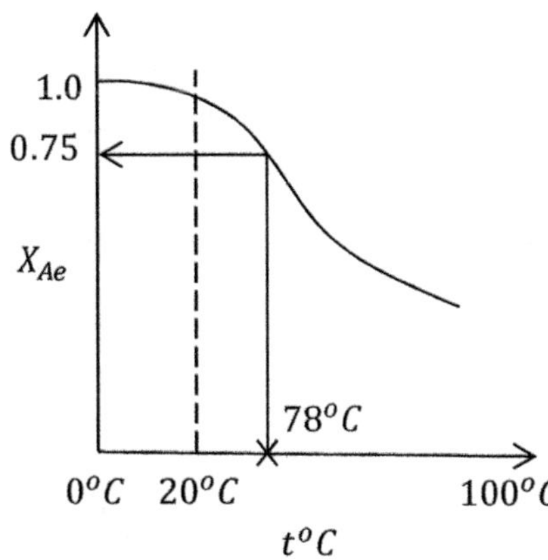

Preparation of rate: Conversion-temperature chart from kinetic data:

(Reversible reaction whether exothermic or endothermic) $A \underset{k_2}{\overset{k_1}{\rightleftharpoons}} B$

$$- r_A = k_1[A] - k_2[B]$$

(1)

$$- \frac{dC_A}{dt} = k_1 C_{Ao}(1 - X_A) - k_2 (C_{Bo} + C_{Ao} X_A)$$

(2) At equilibrium $(r = 0)$:

$$K = \frac{k_1}{k_2} = \frac{C_{Be}}{C_{Ae}} = \frac{X_{Ae}}{1 - X_{Ae}}$$

(1) and (2) gives

$$\frac{k_1 t}{X_{Ae}} = - \ln\left(1 - \frac{X_A}{X_{Ae}}\right)$$

t	C_A	X_A
⋮	⋮	⋮
⋮	⋮	⋮

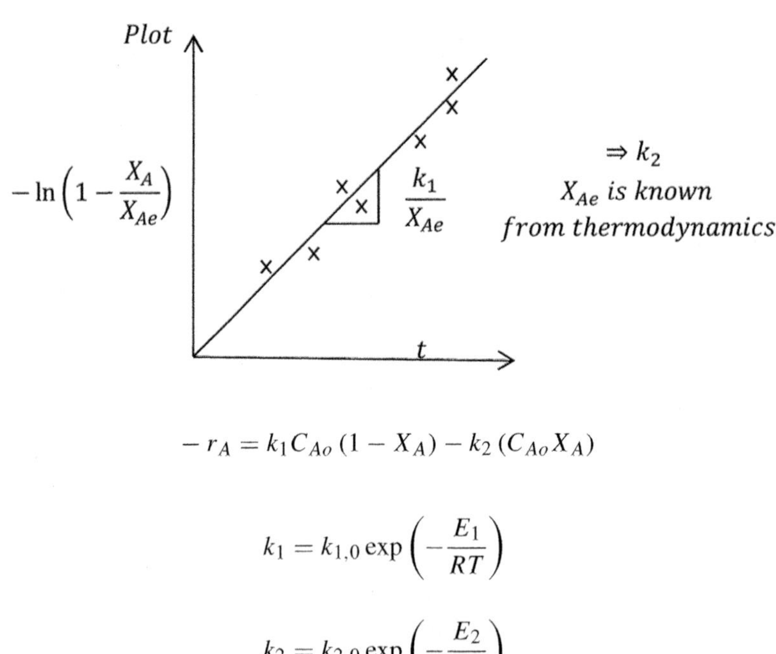

$$- r_A = k_1 C_{Ao}(1 - X_A) - k_2 (C_{Ao} X_A)$$

$$k_1 = k_{1,0} \exp\left(- \frac{E_1}{RT}\right)$$

$$k_2 = k_{2,0} \exp\left(- \frac{E_2}{RT}\right)$$

Carry out reactions at different temperatures

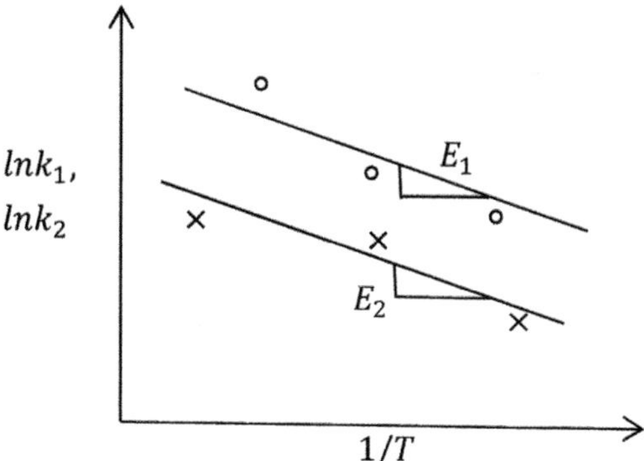

For exothermic reaction $E_1 > E_2$, $\Delta H_r < 0$
For endothermic reaction $E_1 < E_2$, $\Delta H_r > 0$

$$-r_A = k_{10} \exp\left(-\frac{E_1}{RT}\right) C_{Ao}(1 - X_A) - k_{20} \exp\left(-\frac{E_2}{RT}\right) C_{Ao} X_A$$

Step 1:

$$\text{Fix:} \quad -r_A = 0.001, 0.002 \ldots\ldots\ldots\ldots 5 \quad \frac{\text{moles}}{\text{min-lit}}$$

Calculate $X_A(T)$

T	X_A
$-10°C$	⋮
⋮	⋮
110	⋮

Step 2: Plot $X_A(T)$ for different $(-r_A)$

(A) Exothermic:

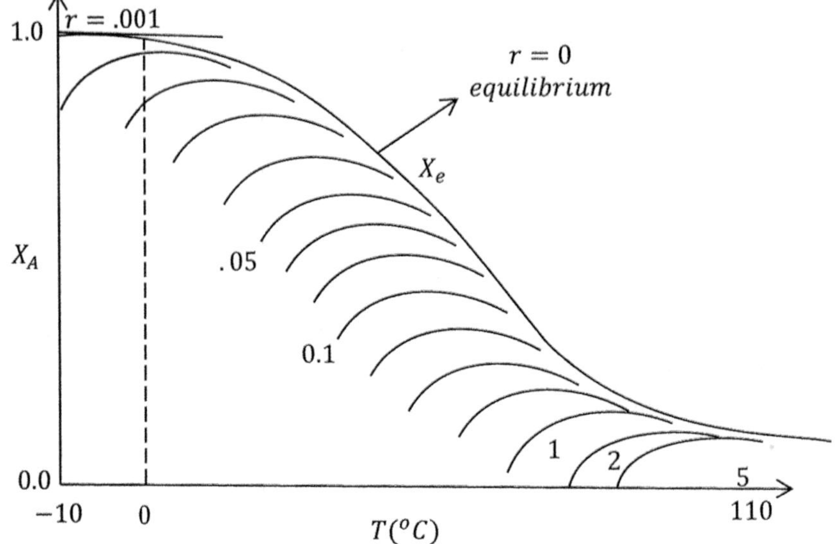

Plot differently:

(1) Fix X_A

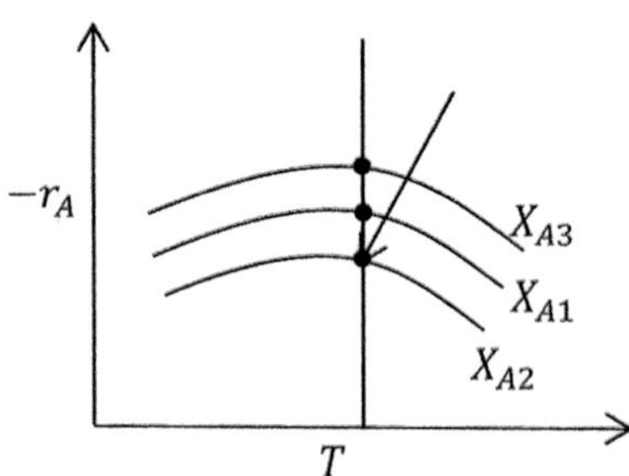

(2) Fix T

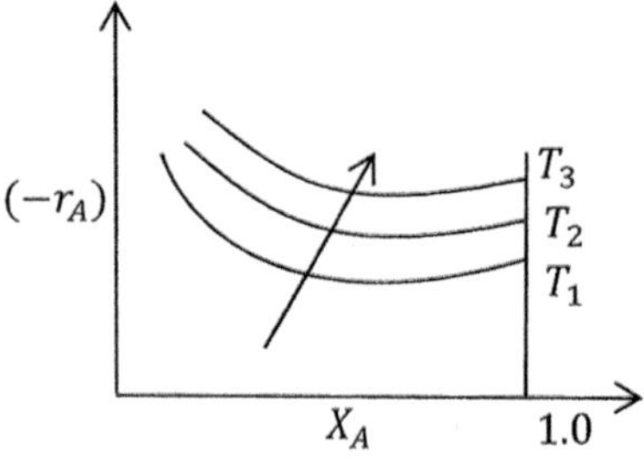

(B) Endothermic ($X_e \uparrow T \downarrow$)

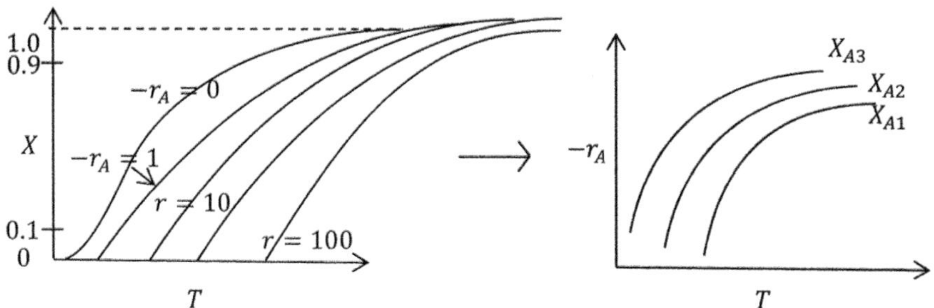

Step 3: Reactor performance characteristics remains the same:

$$F_{Ao}dX = (-r_A)\,dV \qquad\qquad F_{Ao}\Delta X_{Af} = (-r_A)_f\,V$$

$$\tau = C_{Ao}\int_{X_{A1}}^{X_{A2}}\frac{dX_A}{(-r_A)} = ? \qquad\qquad \tau = \frac{C_{Ao}\Delta X_{Af}}{(-r_A)_f} = ?$$

Once

$$- r_A = -r_A(X_A, T)$$

is calculated or plotted, use the specified temperature path in the reactor to determine τ or V of the reactor.

Whether isothermal, exothermic or endothermic reaction, plot $1/(-r_A)$ vs X_A (same as before)

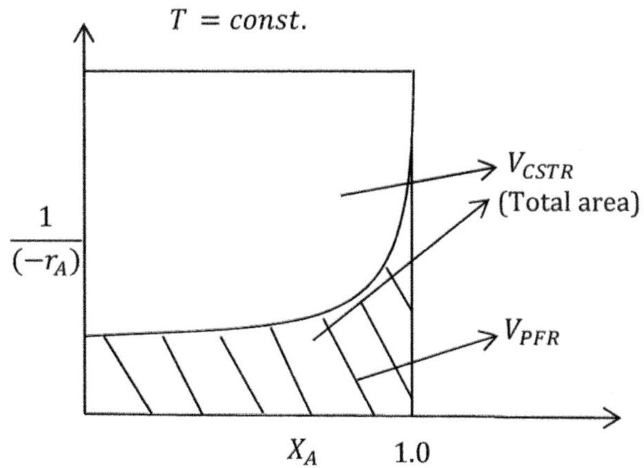

Adiabatic Operation $(\Delta Q = 0)$

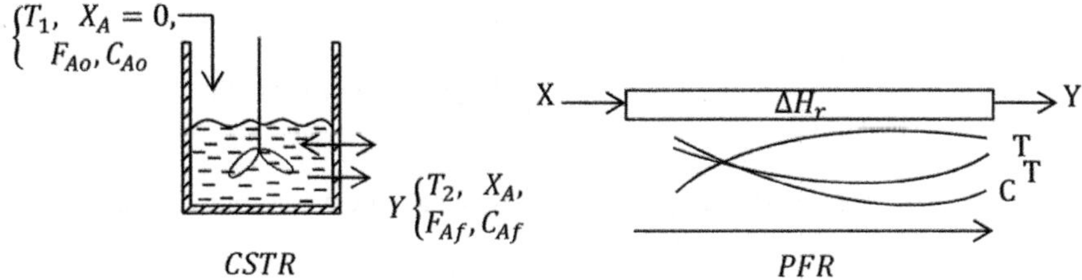

Energy balance under SS (Enthalpy): (choose reference temp $= T_1$)

(Enthalpy in $-$ Enthalpy out $-$ disappearance by reaction $=$ accumulation)$_{\text{rate}}$

CSTR (Consider a single reaction stoichiometry $A \longrightarrow B$)

$$0 - \bar{C}_p'' \left(T_2 - T_1 X_A \right) + \Delta H_{r_1} X_A + \bar{C}_p' \left(T_2 - T_1 \right) \left(1 - X_A \right) = 0$$

Here, C_p' and C_p'' are the mean specific capacity of the unreacted and completely converted product stream per mole of entering reactant A; can be simplified to show that

$$X_A = \frac{\bar{C}_p' \Delta T}{-\Delta H_{r2}}$$

© The Author(s) 2025
N. Verma, *Chemical Reaction Engineering*,
https://doi.org/10.1007/978-3-031-88691-1_15

Energy balance equations:

$$\frac{\Delta X_A}{\Delta T} = \frac{\bar{C}_p}{-\Delta H_r} \; : \; CSTR$$

or

$$\frac{dX_A}{dT} = \frac{\bar{C}_p}{-\Delta H_r} \; : \; PFR$$

(Assuming $C'_p = C''_p$ and the heat of reaction is constant over temperature $T_1 \sim T_2$.)

PFR:

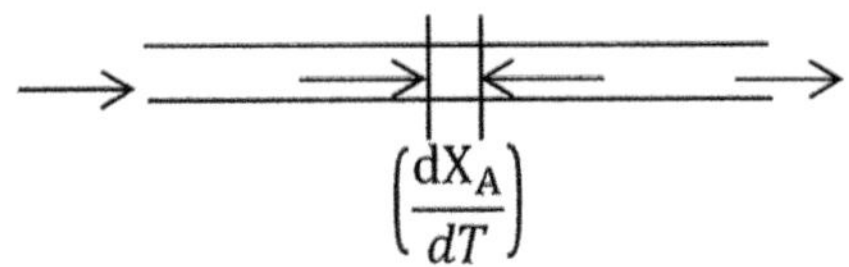

$$\left(\frac{dX_A}{dT}\right)$$

$$\begin{aligned} \text{PFR:} \quad & \frac{dX_A}{dV} = \frac{(-r_A)}{F_{Ao}} \\ \text{CSTR:} \quad & \frac{\Delta X_A}{\Delta V} = \frac{(-r_A)_f}{F_{Ao}} \end{aligned} \Bigg\} \text{ species balance equation}$$

Both of the equations are coupled with energy balance equation to solve for

$$X_{A,T} = f(V)$$

PFR and Adiabatic

$$\frac{dX_A}{dT} = + \frac{\bar{C}_p}{-\Delta H_r}$$

(1) Optimal progression operation (exothermic)

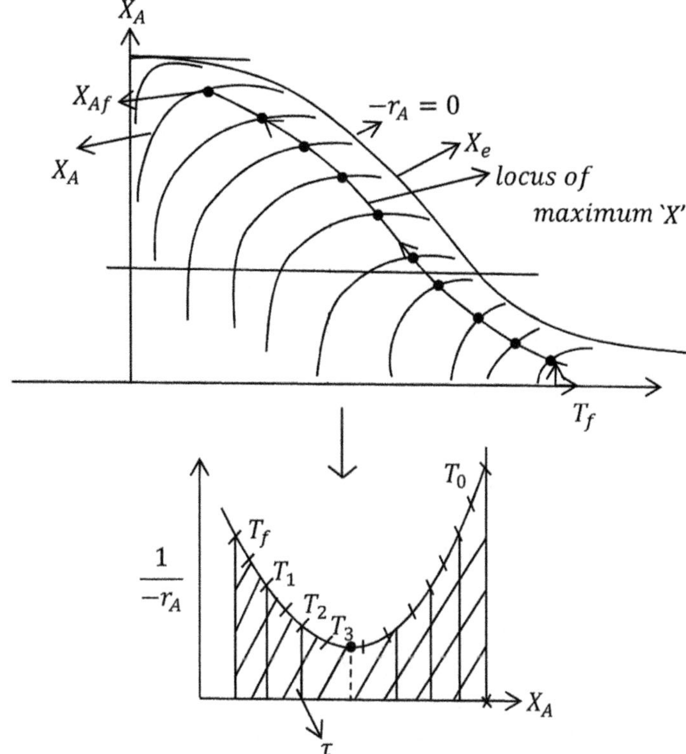

The hatched area or τ corresponding to optimal progression line will give the smallest volume of the reactor. It is only a theoretical number. All other operations will require greater volume.

(2) Adiabatic operation

Endometric ($\Delta H_r > 0$) (adiabatic are $-$ve slope)

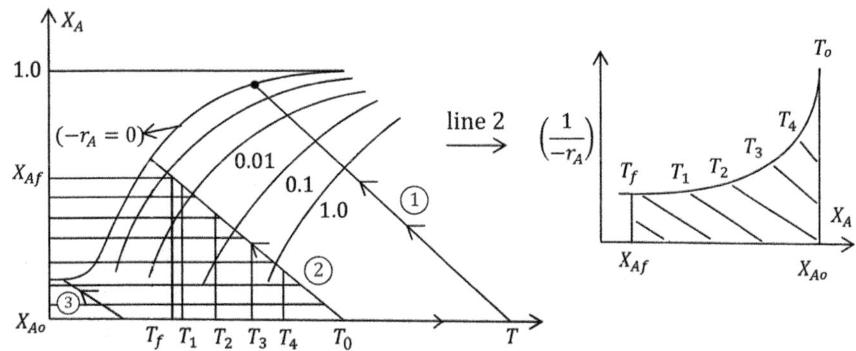

Exothermic ($\Delta H_r < 0$) (adiabatic slopes are $-$ve)

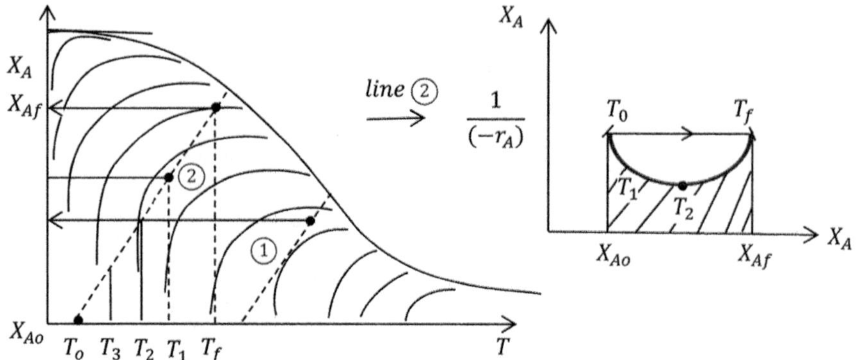

CSTR + Adiabatic (Endothermic ($\Delta H_r > 0$))

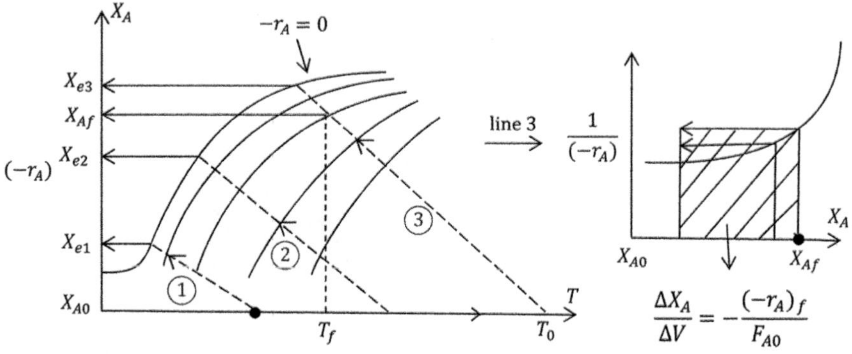

$$\frac{\Delta X_A}{\Delta V} = -\frac{(-r_A)_f}{F_{A0}}$$

(3) Best adiabatic operation: ($\Delta H < 0$)

$$\frac{dX}{dT} = \frac{\overline{C_p}}{-\Delta H_r}$$

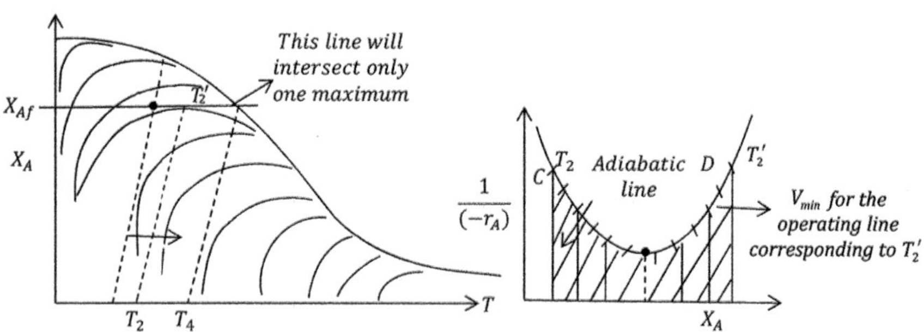

Feed temperature should be adjusted over ($T_2 - T_2'$).

Non-adiabatic Operation
(To allow for heat loss)

$$\frac{dX}{dT} = \frac{c_p - \frac{dQ}{dT}}{-\Delta H_r} \quad \text{(PFR)};$$

$$\frac{\Delta X_A}{\Delta T} = \frac{C_p - \frac{Q}{\Delta T}}{-\Delta H_r} \quad \text{(CSTR)}$$

Q is the heat loss in cal/mol of the reactant.
Note: Adiabatic slope changes.

Multiplicity

(CSTR + Adiabatic)

$$F_{Ao}$$
$$C_{Ao}$$
$$T_o$$

Steady − State (Adiabatic) L

$$(F_{Af}, C_{Af}, T_f);$$

Performance eq^n $\tau = C_{Ao}\dfrac{X_{Af}}{(-r_A)_f}$: *(Species Balance); (τ vs X)*

$$\frac{\Delta T}{X_{Af}} = -\frac{\overline{C}_p}{\Delta H_{r_2}} \quad ; \quad \textit{Energy Balance (T vs X)}$$

Given X_f, what is τ? only one root

$$(X_{Af} \longrightarrow T_f \longrightarrow (-r_A)_f \longrightarrow \tau)$$

But given τ, what is X_{Af}? Multiple roots are possible.
Best way is to take an example.

Example: $A \rightarrow B$ **(CSTR + Adiabatic)**
Given:

$$C_{Ao} = 3 \text{ kmol/m}^3, \quad Q_o = 60 \times 10^{-6} \, m^3/s, \quad \rho_o = 10^3 \frac{kg}{m^3}$$

$$C_p = 4.19 \text{ kJ / kgK}, \quad V = 18 \times 10^{-3} \, m^3, \quad T_o = 298 \, K$$

$$\Delta H_r = -2.09 \times 10^8 \text{ J/ K-mol}$$

Determine X_{Af}

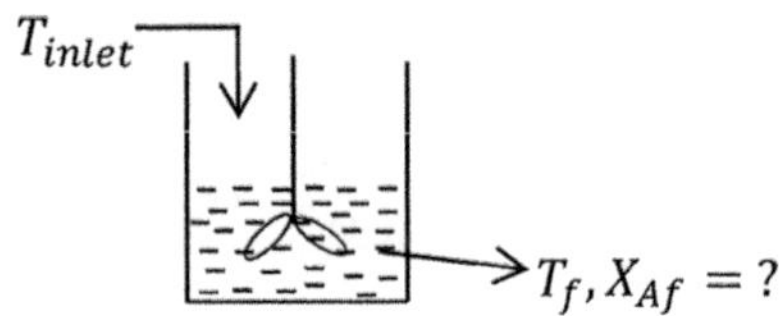

$$-r_A = 4.48 \times 10^6 C_A \exp\left(-\frac{62{,}800}{RT}\right), \frac{kmol}{m^3 - s}$$

Design equations:

$$\tau = C_{Ao}\frac{X_{Af}}{(-r_A)_f} = \frac{V}{V_o} \tag{15.1}$$

Substitute

$$\frac{18 \times 10^{-3}}{60 \times 10^{-6}} = 3 \times 10^3 \times \frac{X_{Af}}{kC_{Ao}\left(1 - X_{Af}\right)}$$

$$= \frac{3 \times 10^3 \times X_{Af}}{4.48 \times 10^6 C_{Ao}\left(1 - X_{Af}\right)\exp\left(-\frac{62{,}800}{RT}\right)} \tag{15.1'}$$

Adiabatic equation:

$$\frac{\Delta X}{\Delta T} = -\frac{\bar{c}_p}{\Delta H_r} \tag{15.2}$$

$$\frac{X_{Af}}{T - 298} = +\left(\frac{4.19 \times 10^3 \frac{kJ}{m^3 K}}{2.09 \times 10^8 \times 3}\right) = \frac{1}{150} \tag{15.2'}$$

$(1')$ and $(2')$ gives

$$\tau = \frac{18 \times 10^{-3}}{60 \times 10^{-6}}$$

$$= \frac{X_{Af}}{4.48 \times 10^6 \left(1 - X_{Af}\right)\exp\left(-\frac{62{,}800}{RT}\right)}$$

$$= 30 \, s$$

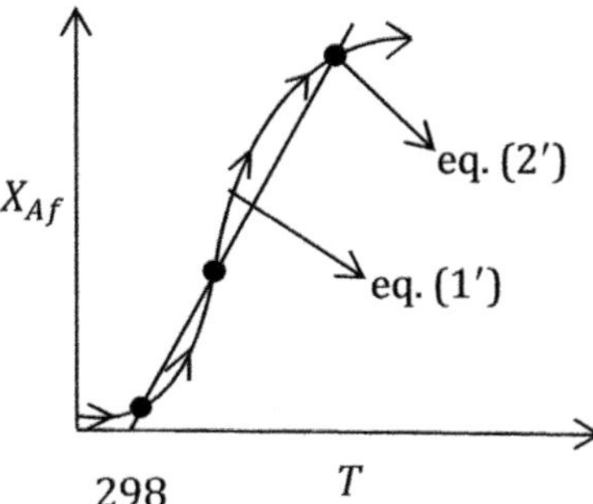

3 roots:

$$X = 1.4\%$$
$$= 31\%$$
$$= 98\%$$

$25 + 3 = 28$ (stable; however, low $T\,\&\,X_{Af}$)

$25 + 48 = 73$ Metastable (ignition point)

$25 + 147 = 172$ (stable; however, high T)

Usually, inlet (feed) temperature is adjusted to avoid "multiplicity", in which case there is only one root. Slope or intercept of the lines $(1' - 2')$ changes by adjusting the feed conditions. (Check for yourself.)

Example 1

A reversible endothermic

$$\left(\dfrac{\Delta H = +\dfrac{20\ \text{kcal}}{\text{mol}}}{\overline{C_p} = 1\dfrac{k\ \text{cal}}{\text{mol-k}}} \right)$$

liquid-phase reaction $A \to B$ is carried out under SS condition in an adiabatic MFR ($F_{Ao} = 1000$ mole/min, $C_{A,in} = 2$ moles/L, $C_{A,exit} = 0.5$ moles/L, $T_{in} = 30°C$). On some day, it was observed that the exit concentration of A had increased to 1 mol/L, while the inlet conditions (temperature and flow rate) were unchanged. The problem was traced to the insulation over the reactor, which was found to be damaged, resulting in heat loss from the reactor to the atmosphere. Determine the rate (cal/min) at which heat was dissipated to the atmosphere.

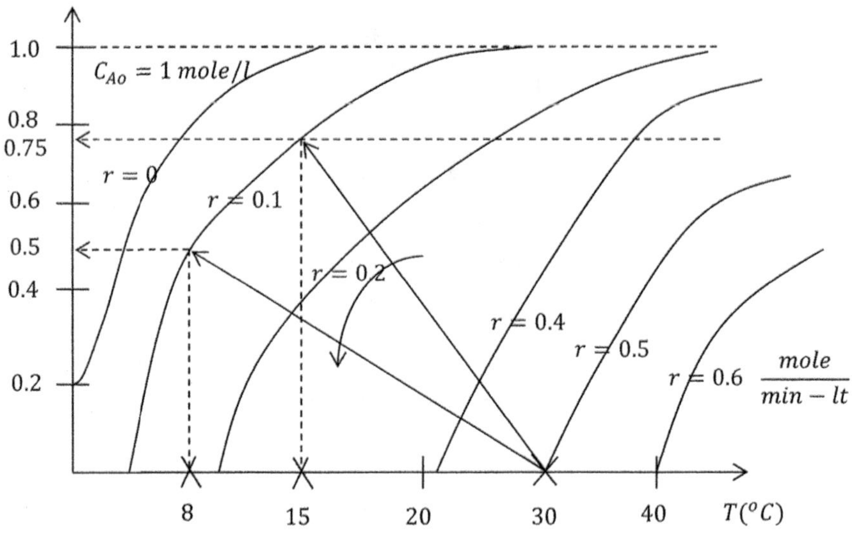

Ans:

Condition 1:

$$X_A = \frac{C_{Ao} - C_{Af}}{C_{Ao}} = \frac{2 - 0.5}{2} = 0.75$$

$$X_A = \frac{\overline{C_p}\Delta T}{(-\Delta H)} \quad \text{(adiabatic line)}$$

$$\text{Slope} = \frac{\overline{C_p}}{(-\Delta H)} = \frac{1000}{-20,000} = -0.05 \left(= -\frac{0.75}{15}\right)$$

Take a slope of

$$\tan\theta = -\frac{0.75}{15} \implies \text{draw the line}$$

This line intersects

$$X_A = 0.75 \implies |\Delta T| = 15°C$$

Alternatively,

$$\Delta T = \frac{-\Delta H X_A}{\overline{C_p}} = \frac{-20.000 \times 0.75}{1000} = -15\,°C$$

$$T_{reactor} \quad \text{or} \quad T_f = 30 - 15 = 15°C$$

Read from graph

$$r = 0.15\frac{\text{mole}}{\text{min-l}}$$

Actual r or $-r_A$ in the reactor $=$

$$2 \times 0.15 = 0.30 \frac{\text{mole}}{\text{min-l}}$$

(Note the plot was given for $C_{Ao} = 1 \frac{\text{mole}}{\text{lt}}$; here, $C_{A,in} = 2$ mole/lt)

$$\tau = \frac{V}{v_o} = \frac{C_{A,in} X_{Af}}{(-r_A)_f} = 2 \times \frac{0.75}{0.30} = 50 \text{ min}$$

$$V = 50 \times \left(\frac{F_{Ao}}{C_{A,in}} \right) = 50 \times \frac{1000}{2} = 2500 \, L$$

Condition 2: Insulation is damaged (heat-loss)

$$X_A = \frac{C_p \Delta T - Q}{(-\Delta H)} \qquad \text{(slope changes)}$$

$$X_{Af} = \frac{2 - 1}{2} = 0.5$$

Also,

$$\tau = \frac{V}{v_o} = C_{A,in} \frac{X_{Af}}{(-r_A)_f}$$

$$(-r_A)_f = \frac{C_{A,in} \times X_{Af} \times v_o}{V} = \frac{1000 \times 0.5}{2500} = 0.2 \frac{\text{mole}}{\text{min-lt}}$$

Actual $(-r_A)_f$ in the graph

$$(\text{for } C_{Ao} = 1 \text{ mol / L}) = 0.1 \frac{\text{mole}}{\text{min-l}}$$

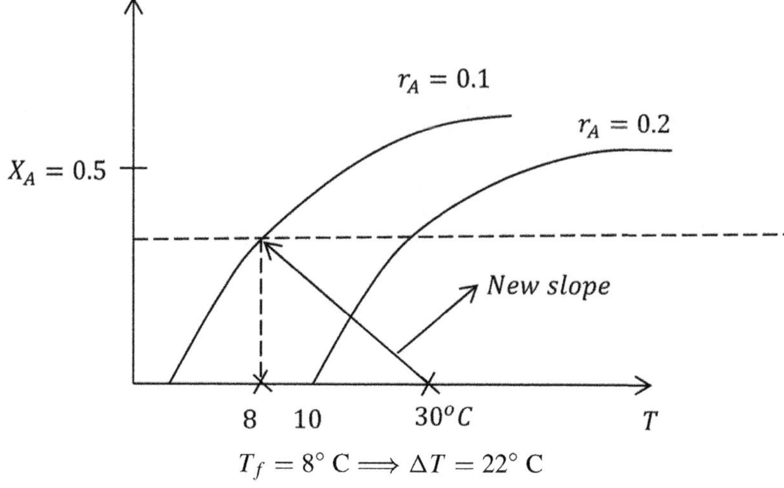

$$T_f = 8^\circ \text{ C} \implies \Delta T = 22^\circ \text{ C}$$

Substituting,

$$0.5 = \frac{1000 \times 22 - Q_{loss}}{(-20,000)}$$

$$Q_{loss} = 22000 + 10000$$
$$= 32,000 \text{ cal / mol}$$
$$= 32 \text{ kcal / mol} = 32 \times F_{Ao} \text{ kcal / min}$$

Example 2

A reversible exothermic, liquid-phase elementary reaction $A \to B$ is carried out in
an ideal MFR ($V = 4200$ L) operated under adiabatic condition to process a feed:
$F_{Ao} = 1200$ moles/min, $C_{Ao} = 2$ moles/L. The temperature of the feed at the inlet
to reactor $= 7°C$, and the concentration of A in the reaction mixture in the reactor
is measured to be 0.6 mol/L under a SS operation. What is the minimum size of an
adiabatic-PFR to process the same feed ($F_{Ao} = 1200$ moles/min, $C_{Ao} = 2$ moles/L)
for the same exit concentration (0.6 mol/L) as in the previous case? For the latter case
(PFR), what is the ratio of heat duties (W) of heat exchangers, placed before and
after the PFR, if the feed is provided at 25°C and the product is delivered also at the
same temperature? Assume constant thermo-physical properties in all calculations.

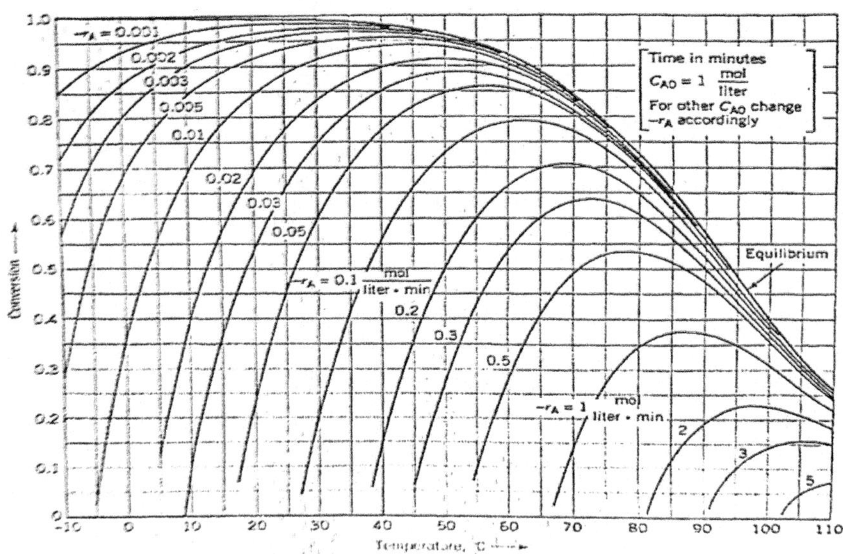

Ans: MFR

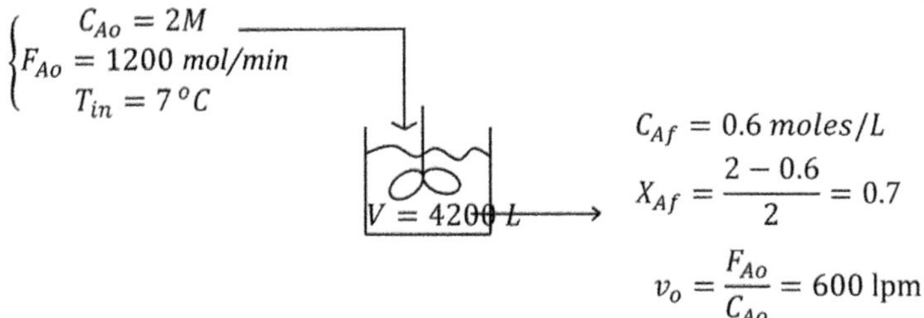

$$\left\{ \begin{array}{l} C_{Ao} = 2M \\ F_{Ao} = 1200 \; mol/min \\ T_{in} = 7\,^{\circ}C \end{array} \right.$$

$$C_{Af} = 0.6 \; moles/L$$

$$X_{Af} = \frac{2 - 0.6}{2} = 0.7$$

$$v_o = \frac{F_{Ao}}{C_{Ao}} = 600 \; \text{lpm}$$

$$\tau = \frac{V}{v_o} = C_{Ao}\frac{\Delta X_{Af}}{(-r_A)_f} \implies \frac{4200}{600} = 7 \; \text{min}$$

$$= 2 \times \frac{0.7}{(-r_A)_f}$$

$$= \frac{1.4}{(-r_A)_f} \implies (-r_A)_f$$

$$= 0.2 \; \frac{\text{moles}}{\text{l-min}}$$

From the plot:

$$\left\{ \begin{array}{l} C_{Ao} = 1 \, \text{moles/L} \\ (-r_A)_f = 0.1 \frac{\text{mole}}{\text{l-min}} \end{array} \right.$$

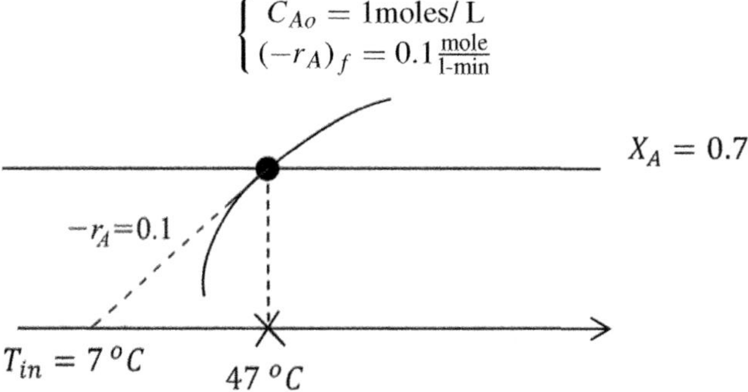

Draw the line from $T_{in} = 7°C$ to the intersection of $X_A = 0.7$ and

$$(-r_A) = 0.1 \implies T_{reactor} = 47°\,C$$

Slope of the adiabatic line

$$\frac{0.7}{47 - 7}(\tan\theta)$$

or

$$\frac{\overline{c_p}}{(-\Delta H)} = \frac{0.7}{40} = 0.0175$$

PFR: (minimum size)

$$\left.\begin{array}{l} F_{Ao} = 1200 \text{ moles / min} \\ C_{Ao} = 2 \text{ moles / L} \\ C_{Af} = 0.6 \text{ moles / L} \end{array}\right\} \text{ Same as before}$$

$$X_{Af} = 0.7 \text{ same as before}$$

Plot:

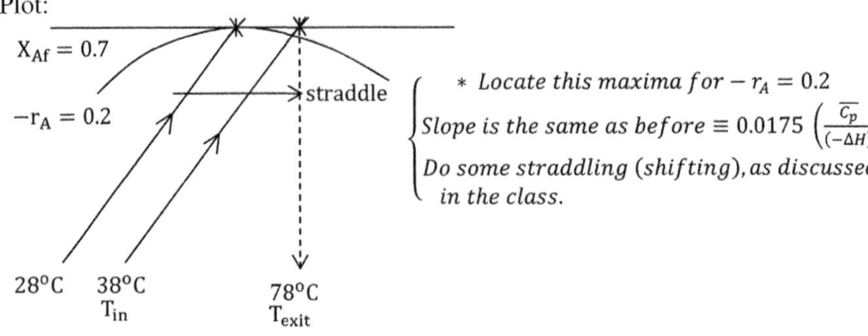

This adiabatic line will intersect $X_A = 0.7$ at $T = 78°C$, which is the reactor temperature or $T_{exit} = 78°C$. The corresponding inlet temperature $T_{in} = 38°C$.

You will have to read a few points on the lines intersecting the adiabatic line between $T_{exit} = 78°C$ for $X_A = 0.7$ and $T_{in} = 38°C$ for $X_A = 0$.

$$\tau_p = \frac{V}{v_o} = C_{Ao} \int_0^{X_{Af}} \frac{dX_A}{(-r_A)}$$

$$= C_{Ao} \sum \frac{\Delta X_A}{(-\bar{r}_A)}$$

X_A	0.7	0.68	0.55	0.3	0.15	0.0
$-r_A$	0.1	0.2	0.4	0.4	0.30	0.25
$1/(-r_A)$	10	5	2.5	2.5	3.33	4
ΔX_A	-	0.01	0.065	0.125	0.075	0.075
$1/[(-r_A)_{avg}]$	-	7.5	3.75	2.5	2.916	3.667

$$V = v_o \times C_{Ao} \text{ (integral)} = 675 \text{ L}$$

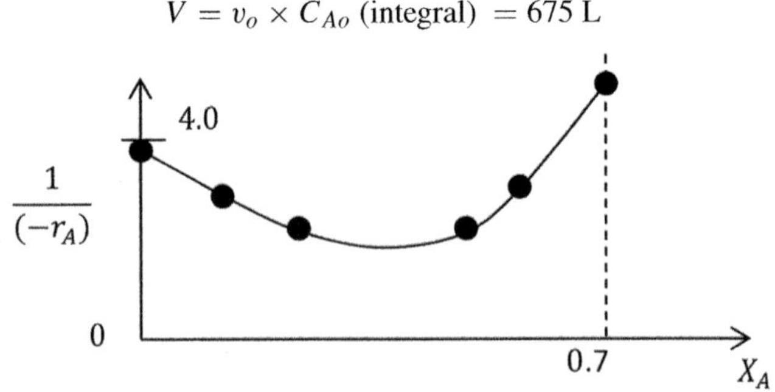

Heat duties:

$$Q_1 \text{ (Pre-heater)} = \dot{m}C_p(T_{in} - T_{feed})$$

$$= \dot{m}C_p(38 - 25)$$

$$Q_2 \text{ (Post reactor)} = \dot{m}C_p\,(T_{reactor} - 25)$$

$$= \dot{m}C_p(78 - 25)$$

$$\text{Ratio} = \frac{Q_1}{Q_2}$$

$$= \frac{38 - 25}{78 - 25}$$

$$= \frac{13}{53}$$

$$= \underline{0.245}$$

Ideal Reactors

Review

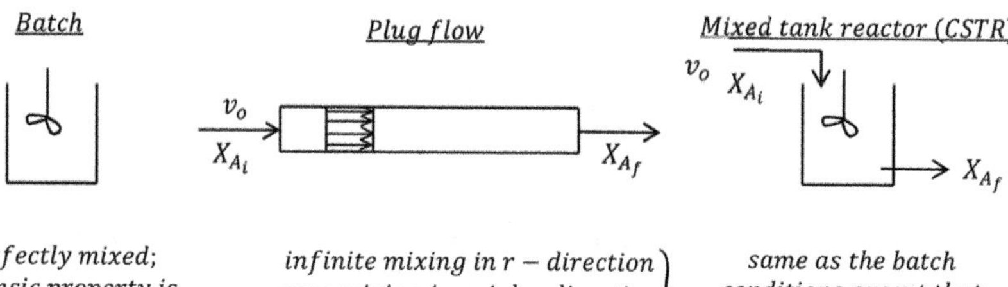

Design Equation of Ideal Reactors

$$v_A A + v_B B + \cdots \rightarrow \text{Products} \ (r \equiv \text{rate of the reaction})$$

Batch

$$t = C_{AO} \int_{X_{A_i}}^{X_{A_f}} \frac{V_o}{V} \frac{dX_A}{(-v_A r)}$$

reaction time
for the conversion to
reach

$$X_A(t) = \frac{N_{AO} - N_A}{N_{AO}}$$

V = volume of the reaction
mixture (t)

X_A = conversion(t)

V_o = volume of the mixture
at reference time

N_{AO} = #of moles at reference time

N_A = #of moles at time 't'

© The Author(s) 2025

N. Verma, *Chemical Reaction Engineering*,

https://doi.org/10.1007/978-3-031-88691-1_16

Plug Flow Reactor

$$\tau = \frac{V}{v_0} = C_{AO} \int_{X_{A_i}}^{X_{A_f}} \frac{dX_A}{(-\nu_A r)}; \quad v_0 = \text{volumetric flow rate of feed at reference point}$$

(space − time)

If flow: $X_A = \left(\dfrac{F_{AO}-F_A}{F_{AO}}\right)$

(# of moles/time) $\equiv$ molar flow rate

(*Note*: space time $\neq$ residense time because of volume-change)

CSTR

$$\tau = \frac{V}{v_0} = \frac{C_{AO}\left(X_{A_f} - X_{A_i}\right)}{(-\nu_A r)_f}$$

design: $\left. \begin{array}{l} \tau\,(\mathbf{X}_A) \\ \text{or } \ X_A(\tau) \end{array} \right\}_?$

If there is no volume—change (i.e., no expansion)

$$t = C_{AO} \int_{X_{A_i}}^{X_{A_f}} \frac{dX_A}{(-\nu_A r)} : \quad \text{batch reactor}$$

$$\tau = C_{AO} \int_{X_{A_i}}^{X_A} \frac{dX_A}{(-\nu_A r)} : \quad \text{plug flow reactor (PFR)}$$

$$t = \tau !$$

batch time = space time

Notes:

(1) If there is no expansion, differential volume in a PFR moves without mixing with the neighboring elements, i.e., each volume element is like a small "batch" reactor with one concentration. Therefore, batch time is the same as space time of an ideal PFR.

V (plug flow) << V (CSTR)

*gradual decrease in   abrupt decrease
the concentration   in the concentration*

(2)

for the same conversion, if order $> +ve$, and *vice-versa*.

Therefore,

$$C_{\text{avg}} \gg C_{\text{exit}} ,$$
$$r_{PFR} \gg r_{CSTR} ,$$
$$V_{PFR} \ll V_{CSTR} ,$$

(3) Without expansion:

$$X_A = \frac{N_{AO} - N_A}{N_{AO}} = \frac{C_{AO} - C_A}{C_{AO}} \quad (V = \text{const.})$$

$$C_A = C_{AO}(1 - X_A)$$

With expansion:

$$X_A = \frac{N_{AO} - N_A}{N_{AO}} : \quad V = V_O(1 + \varepsilon_X X_A)$$

$$C_A = C_{AO}\left(\frac{1 - X_A}{1 + \varepsilon X_A}\right)$$

Multiple Reactors

PFR in Series

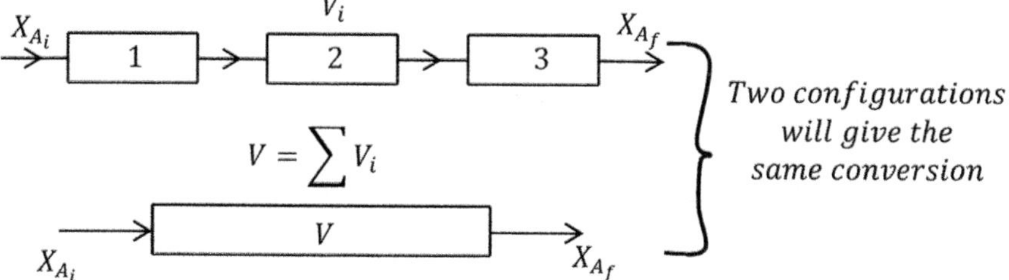

CSTR in Series

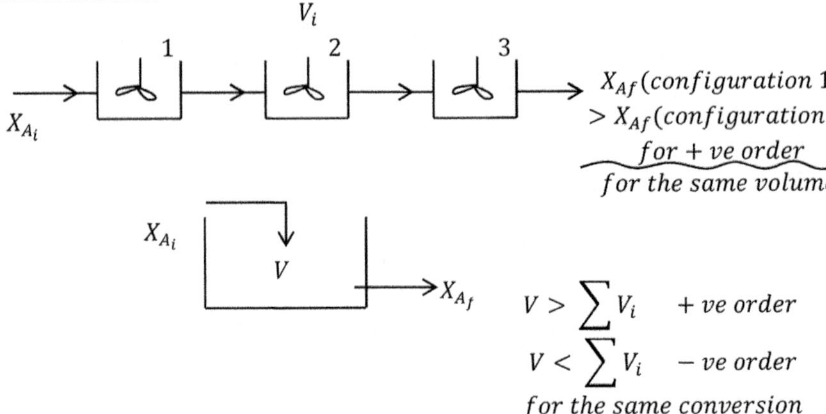

X_{Af} (configuration 1)
$> X_{Af}$ (configuration 2)
for $+\,ve$ order
for the same volume.

$$V > \sum V_i \quad +\,ve\ order$$

$$V < \sum V_i \quad -\,ve\ order$$

for the same conversion

Parallel Configurations

1. PFR:

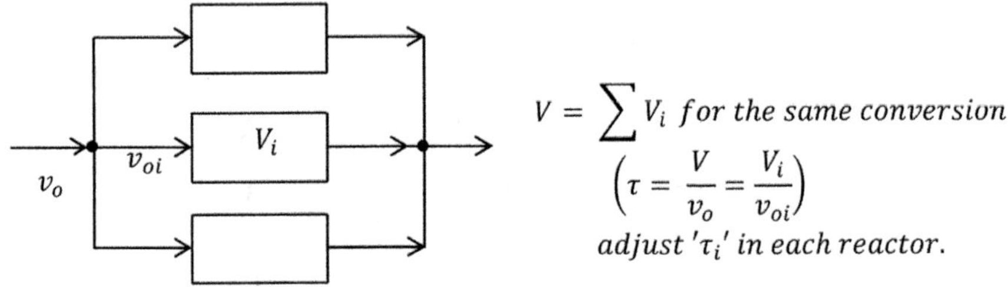

$$V = \sum V_i \ for\ the\ same\ conversion$$

$$\left(\tau = \frac{V}{v_o} = \frac{V_i}{v_{oi}} \right)$$

adjust $'\tau_i'$ in each reactor.

2.

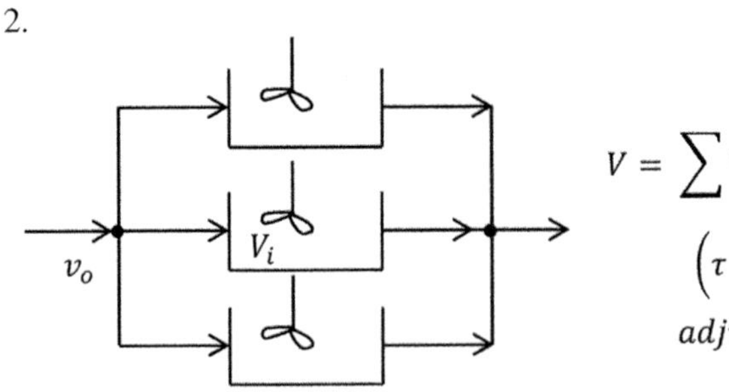

$$V = \sum V_i \ for\ the\ same\ conversion$$

$$\left(\tau = \frac{V}{v_o} = \frac{V_i}{v_{oi}} \right)$$

adjust $'\tau_i'$ in each reactor.

Non-ideal Reactors

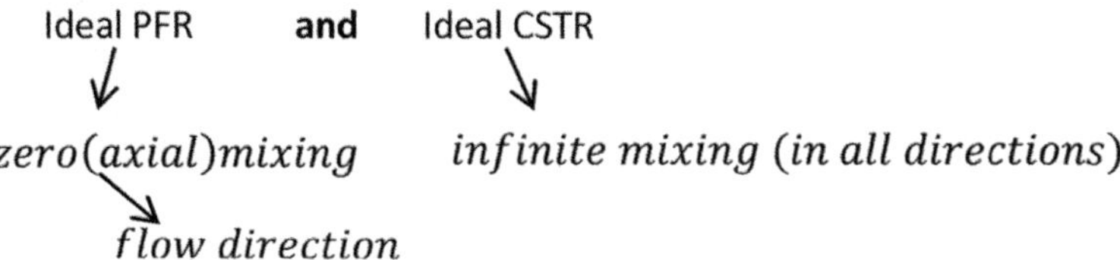

Two extreme situations of ideality, or two model reactors, but infinite mixing in radial direction

Real Reactors
1.

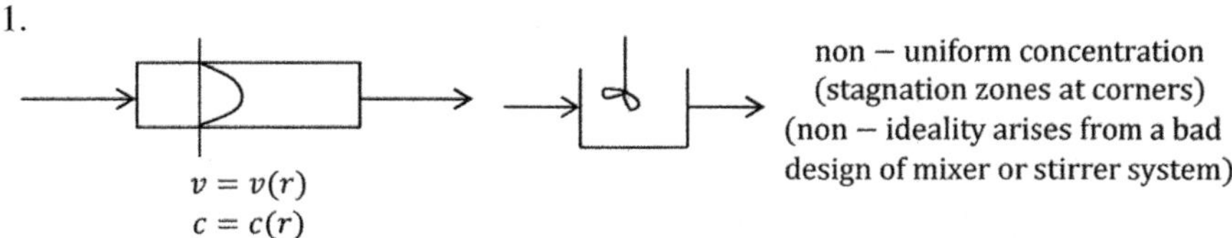

2. Gas flow over packed bed:

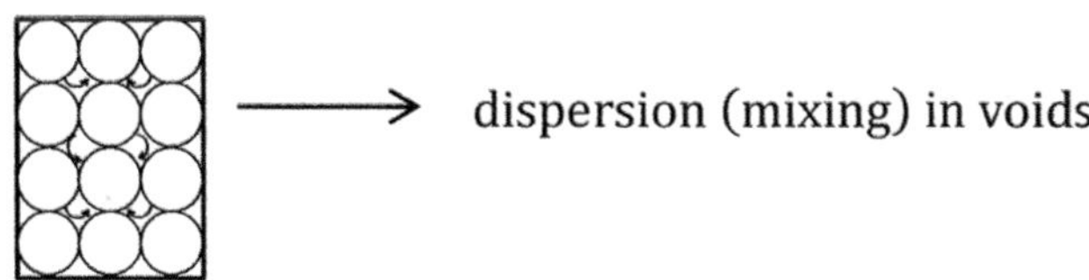

3. Temperature gradient induces convective flow

$$\text{non-isothermal} \longrightarrow \text{mixing}$$

4. $A \longrightarrow R$

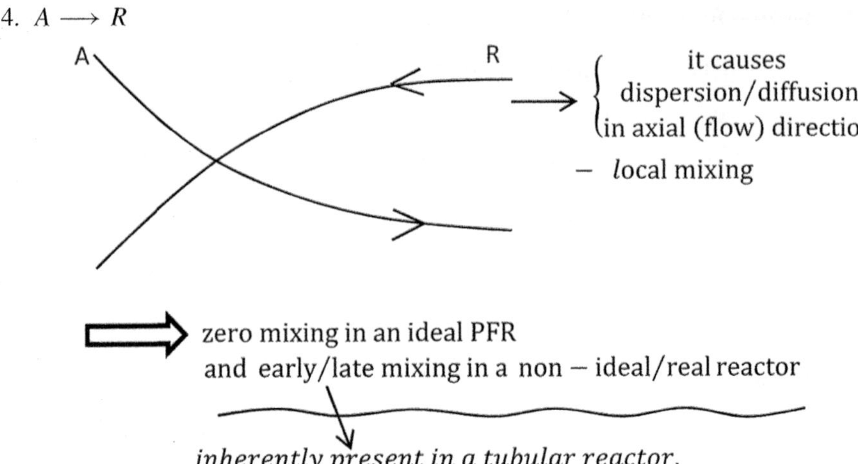

zero mixing in an ideal PFR
and early/late mixing in a non − ideal/real reactor

inherently present in a tubular reactor.

Micro-mixing: Mixing of fluid (elements) on molecular level.

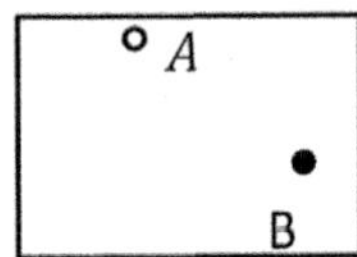

there is a finite probability of
A & B mixing somewhere in the reactor.

Macro-mixing:

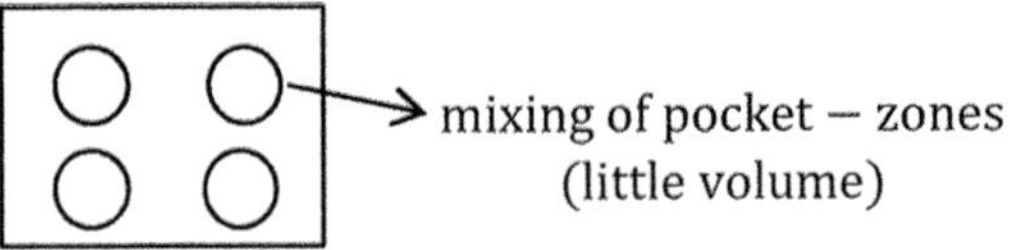

mixing of pocket − zones
(little volume)

Concept of "Pockets" and Criteria for Macro-mixing

1. Contents of individual pockets are well mixed with themselves, but do not mix with second pockets.
2. Pockets should represent chemical reaction conditions. For $A + B \rightarrow C$, one cannot, therefore, reduce the size of a pocket to one molecule; then the reaction will not occur. Pockets should have a finite volume.
3. Pocket contents should have same residence time, but different pockets may have different residence times.

Macro-mixing $\rightarrow$ macro-fluids (segregated fluids) $\rightarrow$ macro-reactor

Ideal PFR is a good example of macro-reactors

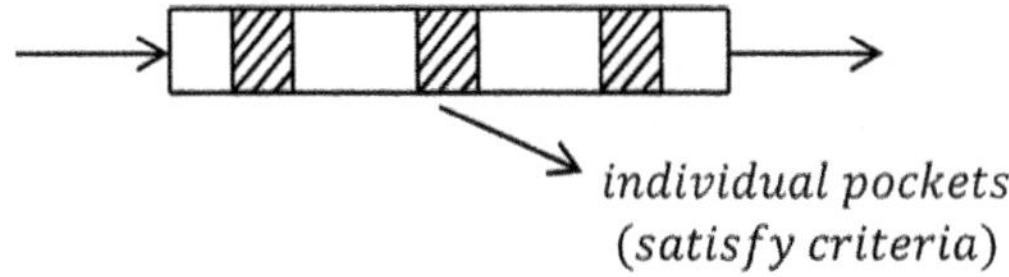

Fast fluidized bed reactor and liquid-liquid emulsion (LLE) membrane-based reactive continuous separation processes are examples of industrial or practical reactors. In the former reactor, particles do not mix with each other and a spatially constant concentration may be assumed within the micron-sized particles. In the second case, small globules (or droplets) are dispersed and immiscible in a continuous phase without mixing.

© The Author(s) 2025

N. Verma, *Chemical Reaction Engineering*,

https://doi.org/10.1007/978-3-031-88691-1_17

Micro-mixing $\rightarrow$ micro-fluids (non-segregated fluids) $\rightarrow$ micro-reactor

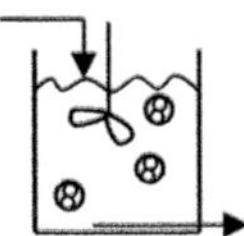

1. Ideal CSTR is a good example of micro-reactors.
2. Ideal CSTR cannot be a macro-fluid reactor because residence times are different for different fluid elements. However, for $A \rightarrow B$ (1st order reaction), an ideal CSTR can be treated as a macro-reactor. In other words, it makes no difference for first-order kinetics whether one assumes non-segregated or segregated flow, or micro- or macro-mixing in an ideal CSTR. Time of reaction alone and not degree of micromixing determines conversion. This is expected because a first-order process depends upon time of reaction of the reacting species and not on the interaction of molecules (See Carberry). In such case, you can think of the size of the pocket as big as the reactor size.

Design of a Non-ideal Flow Reactor ($\epsilon = 0$)

1. Segregated flow model (a real/non-ideal reactor is modeled as a segregated or macro-reactor) V

2.

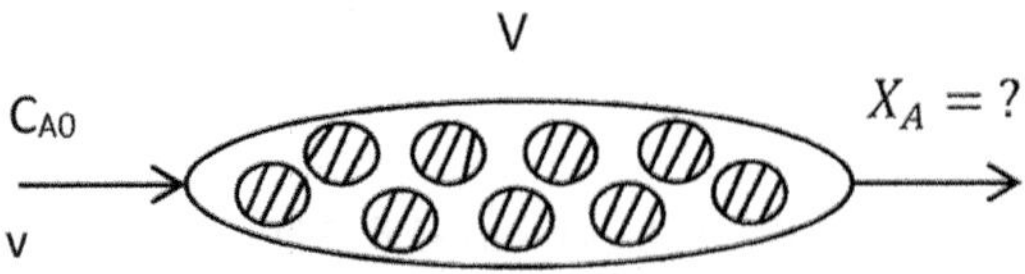

- Each pocket is a batch reactor (well-mixed) within themselves. $X = X(t)$ is known a priori for all types of reaction.
- Each pocket has a residence time (some pockets can exit early and some can exit late).

RTD (We should be able to determine RTD of each pocket or volume-element of the reactor)

Definition (there are two types of RTDs)

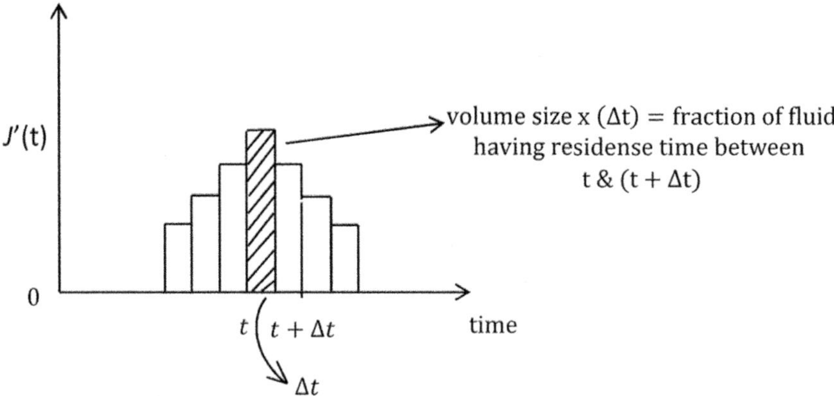

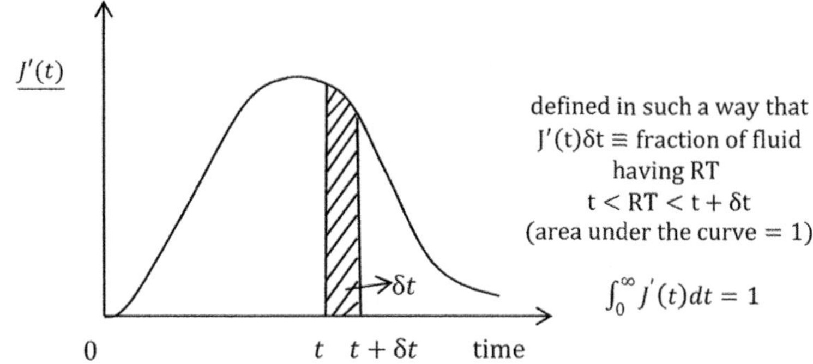

defined in such a way that
J′(t)δt ≡ fraction of fluid
having RT
t < RT < t + δt
(area under the curve = 1)

$$\int_0^\infty J'(t)dt = 1$$

or

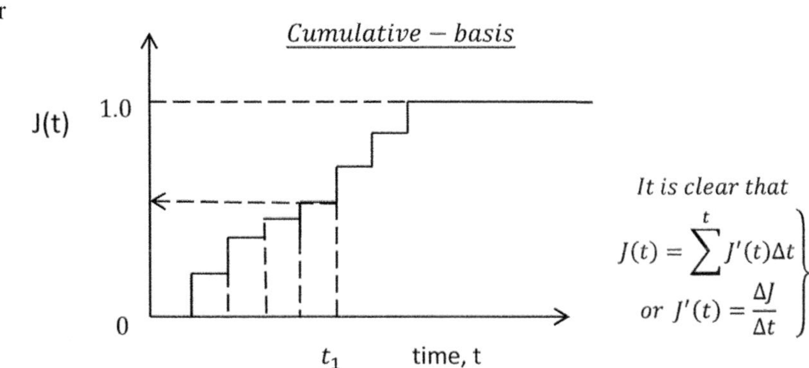

It is clear that

$$J(t) = \sum^{t} J'(t)\Delta t$$

$$or\ J'(t) = \frac{\Delta J}{\Delta t}$$

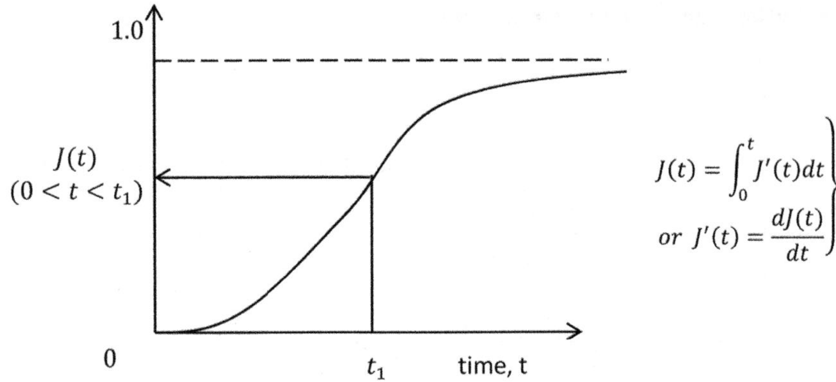

$$J(t) = \int_0^t J'(t)\,dt$$

$$\text{or } J'(t) = \frac{dJ(t)}{dt}$$

Thus, for a non-ideal reactor:

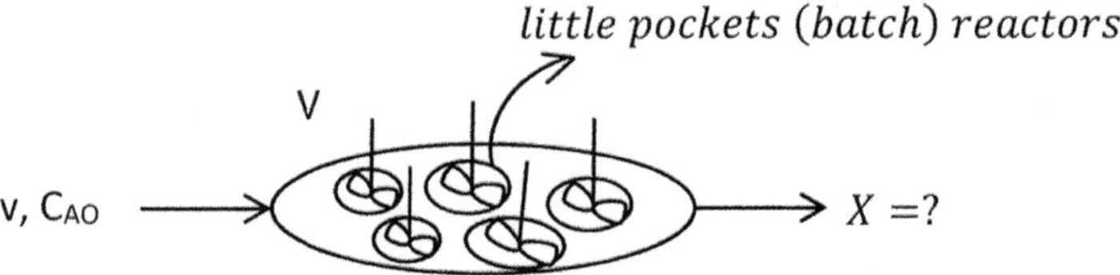

We require the following:

1. $X(t)$ for a batch reactor
2. RTD (J or J') of the reactor under consideration

(*Note*: $t \equiv RTD$ time and $t \neq$ real time)

$$\bar{X} = \underbrace{\int_0^\infty X(t)J'(t)\,dt}_{\text{(averaged over RT)}} \quad \text{or} \quad \sum X(t)J'(t)\Delta t$$

Performance/design equation of a segregated/macro-reactor.

Similarly, any intrinsic property, $\bar{r}$ at outlet

$$= \int_0^\infty r(t)J'(t)\,dt$$

where $r(t)$ is the *intrinsic property* and $J'(t)$ is the *fraction of fluid having RT between t & $t + \Delta t$*

The average residence time always holds good as follows:

$$(\tau) \text{ or } \bar{t} = \overline{RTD} = \int_0^\infty t\,J'(t)\,dt$$

Determining RTD Experimentally (Two Methods)
(tracer analysis: no reaction)

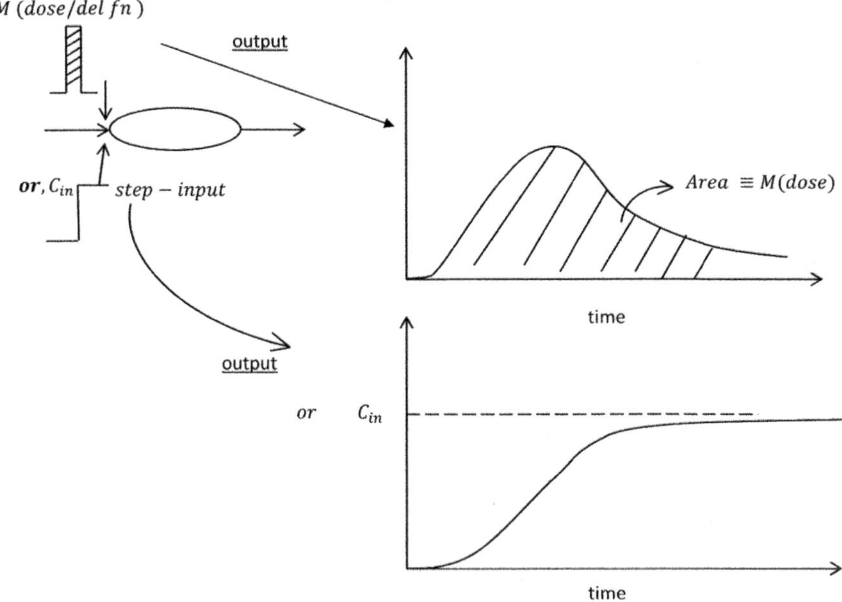

Either of the two methods of injecting a tracer will work. It is a question of convenience in the experiment. "Q", flow rate must be the same as that used in the real reactor; however, without reaction (i.e., reaction is switched off!)

Method 1: Step Tracer Injection

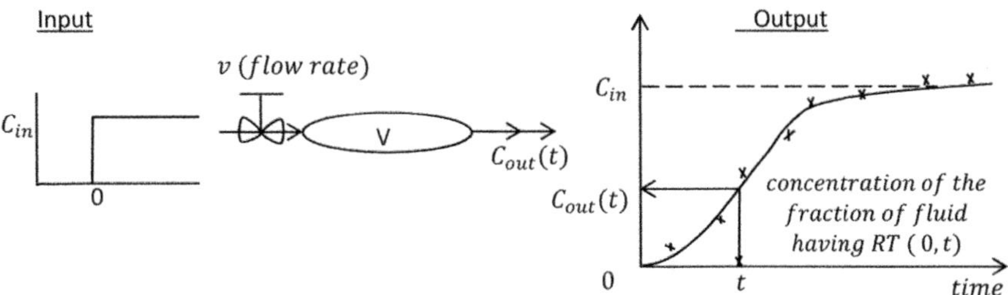

Calculations: By the very definition of $J(t)$:

$$J(t) = \frac{C_{\text{out}}(t)}{C_{\text{in}}} \quad \Rightarrow \quad J'(t) = d(J(t))/dt$$

Now, turn on the reaction (no tracer!)

$$\bar{X} = \int_0^\infty \underset{\text{Batch}}{X(t)}\, J'(t)\,dt = \int_0^\infty X(t)\,d(J(t))$$

Method 2: Pulse Injection

Input

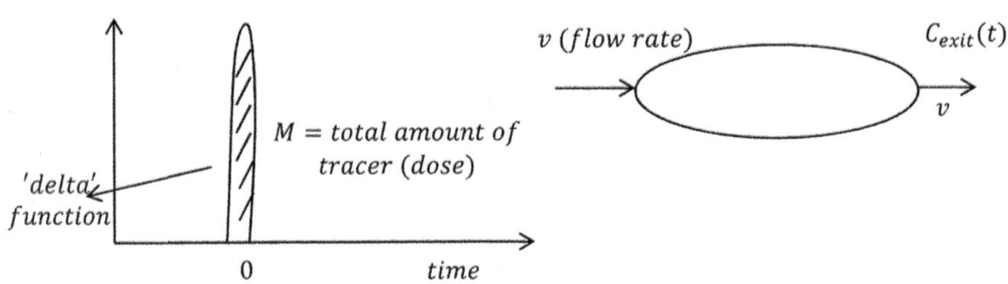

Output

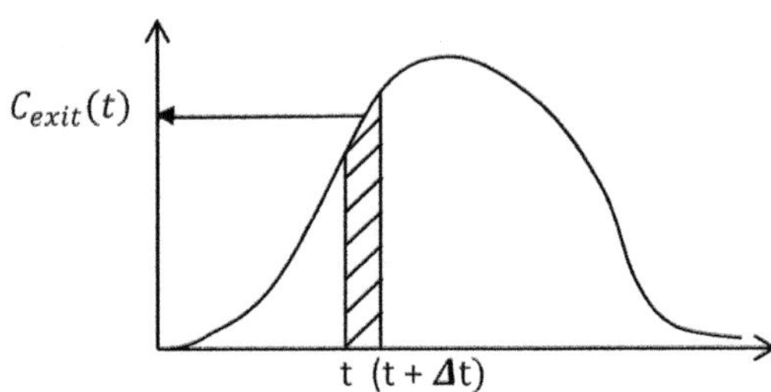

Species balance (use definition of $J'(t)$)

$$M J'(t)dt = C_{\text{exit}}(t)v\,dt$$

$$J'(t) = \frac{C_{\text{exit}}\,v}{M} = \frac{C_{\text{exit}}(t)}{M/v}$$

$$= \frac{C_{\text{exit}}(t)}{\int_0^\infty v C_{\text{exit}}(t)dt/v} = \frac{C_{\text{exit}}(t)}{\int_0^\infty C_{\text{exit}}\,dt}$$

$$J(t) = \int_0^t J'(t)dt$$

Again, turn on the reaction (no tracer)

$$\bar{X} = \int_0^\infty X(t)J'(t)dt = \int_0^\infty X(t)d(J(t))$$

Either of the two methods will work: by determining $J(t)$, $J'(t)$ can be determined, or *vice versa*.

Also, be careful in evaluating or calculating such integrals:

$$\int_0^\infty X(t)J'(t)dt$$

which contain two different functions that may span over different time domains, thus having different limits of integrations. See the examples (and questions in the examinations) that follow in the next lectures.

Segregated reactors (let us validate the segregated model with the reactors of known mixing)

Example 1: Plug-flow reactor

Performance equation

$$X_A = 0$$

$$X \quad X + dX \qquad \rightarrow X_{Af}$$

First-order reaction: $X_A = 1 - e^{-k\tau}$, where τ is the residence time.

$$\left(\text{Performance equation: } \tau = \frac{V}{v_o} = C_{AO} \int_0^{X_A} \frac{dX_A}{(-v_A r)} \right)$$

Recall previous discussion on RTD:

In general   For a PFR, RTD:

$$U(t) \rightarrow \boxed{\quad} \rightarrow J(t)$$

$$J(t) = U(t - \tau)$$

$$\begin{cases} 0 \; ; & t < \tau \\ 1 \; ; & t \ge \tau \end{cases}$$

$$\delta(t) \rightarrow J'(t)$$

$$J'(t) = \delta(t - \tau)$$

$$\begin{cases} 0 \; ; & t \ne \tau \\ \infty \; ; & t = \tau \end{cases}$$

© The Author(s) 2025

N. Verma, *Chemical Reaction Engineering*,

https://doi.org/10.1007/978-3-031-88691-1_18

Let us apply segregated reactor model to a PFR:

$$X_{Af} = \int_0^\infty X(t)J'(t)dt = \int_0^\infty \left(1 - e^{-kt}\right)\delta(t - \tau)dt$$

$$\underset{\text{Batch}}{\uparrow} \qquad = 1 - e^{-k\tau} \quad (X_{Af}(\tau), \text{ same as before})$$

Example 2: CSTR (let us determine RTD first)

$$X_A = 0 \xrightarrow{\quad v \quad} \boxed{\qquad} \xrightarrow{\quad} X_{Af} \quad c(t)$$

$$V$$

Input:

$$M = trace \ amount \ of \ tracer \quad \Rightarrow J'(t)$$

$$\frac{1/\tau}{0 \quad \delta t \quad t}$$

$$J'(t) \equiv \frac{C(t)}{\int_0^\infty C(t)dt} \quad \Rightarrow \quad \text{species (tracer) balance at } (t - t + \Delta t)$$

$$0 - vC(t)dt = \underbrace{VdC}_{\text{in CSTR}}$$

$$\underset{\text{input}}{\uparrow} \quad \underset{\text{output}}{\uparrow}$$

$$\int_{C_o}^{C} \frac{dC}{C} = - \int_0^t \frac{v}{V}dt \Rightarrow \ln\frac{C}{C_o} = -\frac{v}{V}t = -t/\tau$$

$$C(t) = C_o e^{-t/\tau};$$

$$C_o = M/V$$

$$C(t) = M/V e^{-t/\tau}$$

$$J'(t) = \frac{e^{-t/\tau}}{\int_0^\infty e^{-t/\tau}dt}$$

$$= -\frac{e^{-t/\tau}}{\tau(0-1)} = \frac{e^{-t/\tau}}{\tau}$$

$$J(t) = \int_0^t J'(t)dt = 1 - e^{-t/\tau}$$

else, start from the species balance for the step input case:

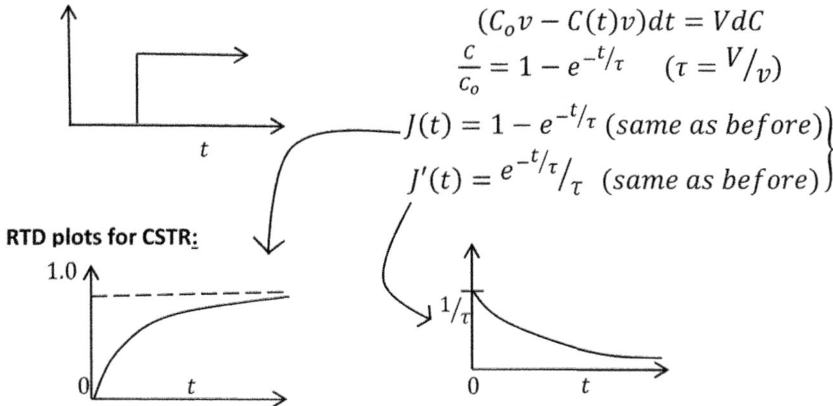

Get back to the performance equation of the segregated reactor:

$$X_{Af} = \int_0^{\infty} X(t)\,J'(t)\,dt$$
$$\uparrow$$
$$\text{batch}$$

Note: CSTR is a micro (non-segregated) reactor; yet the CSTR reactor can be treated for the first-order reaction.

Example 3: Determine conversion for a first-order reaction in an ideal CSTR.

CSTR:

$$X_{Af} = \int_0^{\infty} \left(1 - e^{-kt}\right) \frac{e^{-t/\tau}}{\tau}\,dt$$
$$= \frac{k\tau}{1 + k\tau}$$

same as that from the performance equation.

Recall the design/performance equation:

$$\tau = \frac{v}{v_o} = \frac{C_{Af}}{(-r_A)_f} = \frac{C_{Ao}(X_{Af} - X_{Ai})}{(-v_A r)_f} = \frac{C_{Ao}X_{Af}}{kC_{Ao}\left(1 - X_{Af}\right)}$$
$$= \frac{X_{Af}}{k\left(1 - X_{Af}\right)} \Rightarrow X_{Af} = \frac{k\tau}{1 + k\tau} \quad \text{(same as before)}$$

Example 4: Laminar flow reactor (no diffusion in r direction)

It is a segregated or macro-reactor.

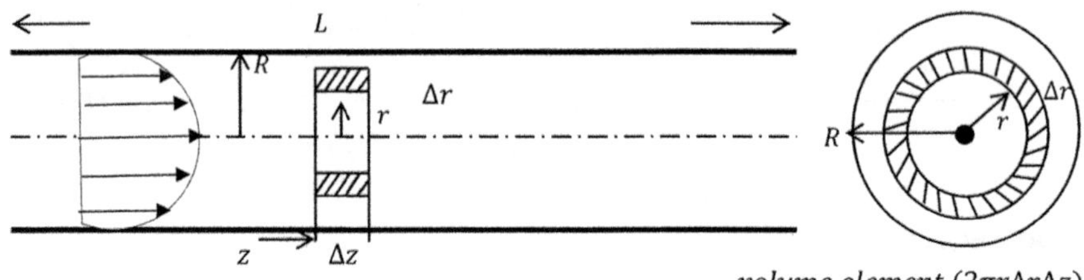

volume element $(2\pi r \Delta r \Delta z)$

Determine $J(t)$ and $J'(t)$ for

$$v = v_o \left(1 - \frac{r^2}{R^2} \right)$$

where v_o is maximum (centerline) velocity. (See the textbooks for derivation)

$$J'(t)$$

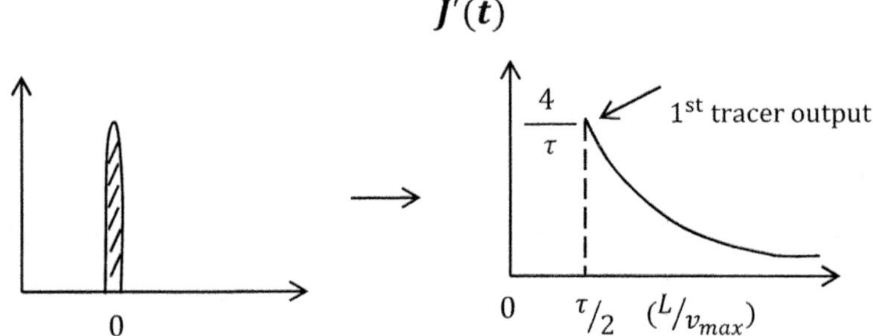

Therefore,

$$J(t) = \int_{\tau/2}^{t} J'(t)dt = 1 - \frac{\tau^2}{4t^2}; \quad J'(t) = \frac{\tau^2}{2t^3}$$

The tracer corresponding to the fluid element at the center of the tube will come out first.

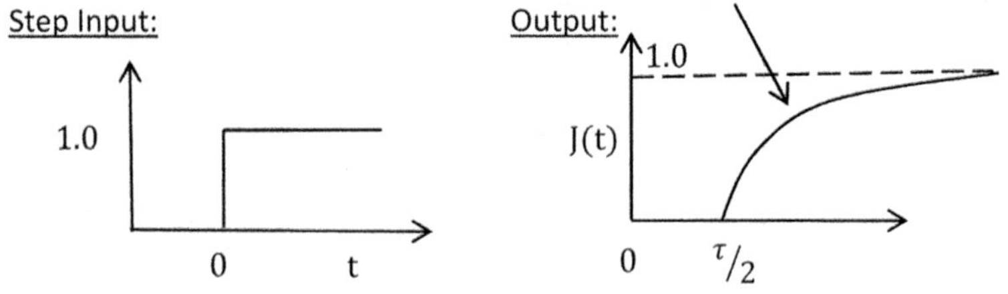

Note:

$$\left. \begin{array}{ll} t = 0 & \text{for } CSTR \\ = \tau & \text{for } PFR \\ = \tau/2 & \text{for } LFR \end{array} \right\} \begin{array}{l} \text{when the first tracer} \\ \text{comes out/exits the} \\ \text{reactor.} \end{array}$$

Plot all three together (qualitatively):

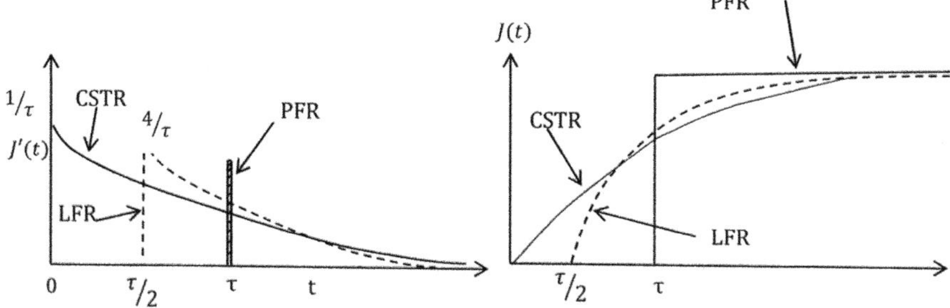

For LFR

$$X = \int_{\tau/2}^{\infty} X(t)\,J'(t)\,dt \quad \left(\text{no tracer until } \frac{\tau}{2} \text{ for LFR}\right)$$

Batch

$$= \int_{\tau/2}^{\infty} (1 - e^{-kt})\frac{\tau^2}{2t^3}\,dt$$

1st order

Similarly, conversions can be calculated for the other rate orders as well.

Two results (RTD segregated model and traditional species balance/ performance equation) will give the same results. See more on this in the book "Chemical & Catalytic Reaction Engineering" by James Carberry.

Example 5:

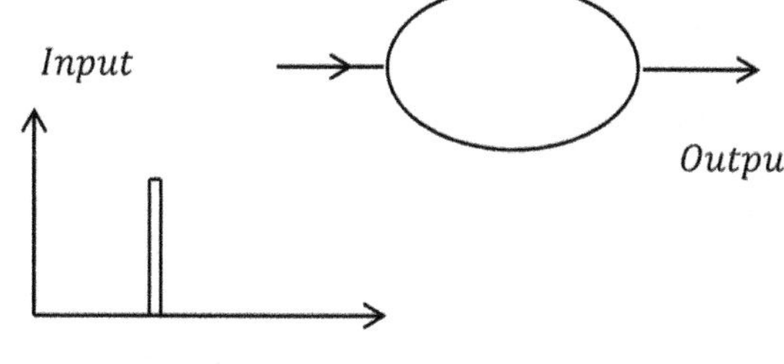

$t(min)$	$tracer(g/L)$
0	0
5	3
10	5
15	5
20	4
25	2
30	1
35	~ 0

Determine $\bar{X}$ and $\bar{\tau}$ ($k = 1\,\text{min}^{-1}$)

Ans:

$$\bar{X} = \int_0^\infty X(t)J'(t)dt \quad : segregated\ reactor$$

$$\tau = \int_0^\infty tJ'(t)dt \quad batch\ \text{kinetics}$$

$k = 1\,\text{min}^{-1} \Rightarrow$ it is a first-order reaction

$$X(t) \equiv \text{batch} = 1 - e^{-kt} = 1 - e^{-t}$$

© The Author(s) 2025

N. Verma, *Chemical Reaction Engineering,*

https://doi.org/10.1007/978-3-031-88691-1_19

$J'(t)$ to be determined from the table below; work on t_{avg} and C_{avg} for an improved numerical accuracy.

t	C	C_{avg}	t_{avg}	$(C\Delta t)$	$J'(t)$ $= C/\sum C\Delta t$	$X(t)$	$X(t)J'(t)\Delta t$
0	0	–	–				
5	3	1.5	2.5				
10	5	4	7.5				
15	5	5	12.5	Calculate	Calculate	Calculate	Calculate
20	4	4.5	17.5				
25	2	3	22.5				
30	1	1.5	27.5				
35	0	0.5	32.5				
				$\sum C\Delta t$ $\downarrow$ 5 min		$\bar{X} =$ $\dfrac{\sum X(t)}{J'(t)\Delta t}$	

Recall:

$$\bar{X} = \int_0^\infty X(t)J'(t)dt$$

holds good for segregated reactors. However, if the rate of reaction is first-order, the reactor can still be treated as a macro-reactor, and the expression holds good. Correct answers are 99.37% conversion and 15 min of average residence time. For the latter calculation, add one more column for $t J'(t) \Delta t$ and add the row elements.

Example 6: A first-order reaction $(A \xrightarrow{k=2\,\text{min}^{-1}} R)$ takes place in a non-ideal mixed reactor. Because of inadequate mixing, only 85% of the reactor volume is well mixed and has the mean residence time of 1 min. The RTD of the remaining 15% (in the corners and close to the walls) can be approximated by the following equation:

$$J(\theta) = 1 - e^{-0.5\theta}\,(0 \le \theta \le \infty)$$

Determine:

(a) The RTD of the reactor

(b) τ, and

(c) $\bar{X}$

Ans:

$J_1(\theta)$ for 85% $V = 1 - e^{-\theta/1}$

$= 1 - e^{-\theta}$ $(\tau_1 = 1)$

$J_2(\theta)$ for 15% $V = 1 - e^{-0.5\theta}$

$(\tau_2 = 2)$

Note that 15% of CSTR volume also behaves as an ideal CSTR, as seen by the corresponding $J(\theta)$

$$J(\theta) = \frac{Q_1 J_1(\theta) + Q_2 J_2(\theta)}{Q_1 + Q_2}$$

$$Q_1 = \frac{0.85\,V}{\tau_1} = 0.85\,V$$

$$Q_2 = \frac{0.15\,V}{\tau_2} = 0.075\,V$$

$$J(\theta) = \frac{0.85 \times \left(1 - e^{-\theta}\right) + 0.075 \times \left(1 - e^{-0.5\theta}\right)}{0.85 + 0.075}$$

$$= \frac{0.925 - 0.85e^{-\theta} - 0.075e^{-0.5\theta}}{0.925}$$

(a)

$$J(\theta) = 1 - 0.919e^{-\theta} - 0.081e^{-0.5\theta}$$

(b)

$$\tau = \frac{V}{Q} = \frac{V}{(0.85 + 0.075)V} = 1.081\ \text{min}\ (def^{n})$$

Alternatively,

$$\tau = \int_0^\infty \theta J'(\theta)d\theta = \int_0^\infty \theta \left(0.919e^{-\theta} + 0.041e^{-0.5\theta}\right) d\theta$$

$$= \left[\theta \left(-0.919e^{-\theta} - \frac{0.041e^{-0.5\theta}}{0.5}\right)\right]_0^\infty$$

$$+ \int_0^\infty \left(0.919e^{-\theta} + \frac{0.041e^{-0.5\theta}}{0.5}\right) d\theta$$

$$= 0 + \left[-0.919e^{-\theta} - \frac{0.041e^{-0.5\theta}}{0.25}\right]_0^\infty$$

$$= \left[-\left(0.919e^{-\theta} + 0.162e^{-0.5\theta}\right)\right]_0^\infty$$

$$= 0.919 + 0.162 = 1.081\ \text{min (same as before)}$$

(c)

$$\overline{X} = \int_0^\infty X(\theta)J'(\theta)d\theta = \int_0^\infty (1 - e^{-k\theta})J'(\theta)d\theta$$

$$\text{(segregated reactor)} \qquad \longrightarrow\ 1^{st}\ order/batch\ reactor$$

$$= \int_0^\infty (1 - e^{-k\theta})(0.919e^{-\theta} + 0.041e^{-0.5\theta})d\theta = \underline{0.677}$$

$$= \int_0^\infty \left(1 - e^{-k\theta}\right) \left(0.919e^{-\theta} + 0.041e^{-0.5\theta}\right) d\theta = \underline{0.677}$$

This must also be the same as calculated below:

As per the problem statement, the real reactor is a combination of two ideal CSTRs in parallel:

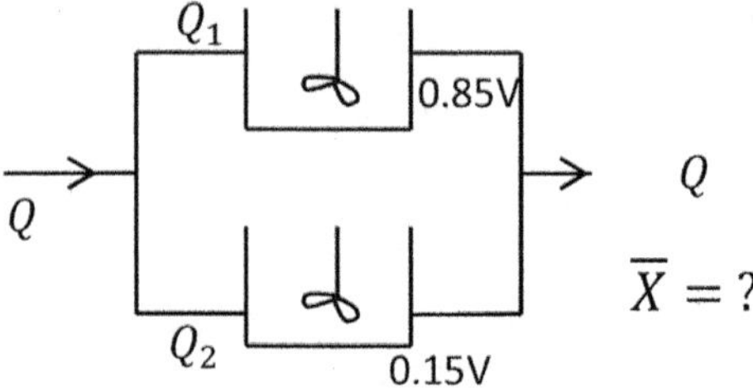

Therefore, X_1 for a first-order rate reactor in an ideal CSTR

$$X_1 = \frac{k\tau_1}{1 + k\tau_1} \quad \text{(performance equation)}$$

$$= \frac{2 \times 1}{1 + 2 \times 1} = \frac{2}{3}$$

$$X_2 = \frac{k\tau_2}{1 + k\tau_2} = \frac{2 \times 2}{1 + 2 \times 2} = \frac{4}{5}$$

Therefore,

$$\overline{X} = \frac{X_1 Q_1 + X_2 Q_2}{Q_1 + Q_2} = \frac{2/3 \times 0.85V + 4/5 \times 0.075V}{0.85V + 0.075V}$$

$$= \frac{0.567 + 0.06}{0.925} = \underline{0.677}$$

(Thus, RTD calculation/data produce the same result as that from design equation)

Example 7: Coal particles are devolatized in a fast fluidized/moving bed reactor at $450°C$, using N_2 gas. The average residence time of the well-mixed particles, $\tau = 10\,\text{min}$. Initial volatile contents are $46\%(w/w)$. The devolatization rate follows first-order rate kinetics:

$$\frac{dX}{dt} = -k(X - 0.2); \quad X \equiv \text{mass fraction of volatiles.}$$

Determine $\overline{X}$ (the average volatile content of the coal particles) at the reactor-exit. Assume $k = 1.0\,\text{min}^{-1}$.

(*Note:* 20% of the volatiles never comes out. That is, the devolatization rate is zero at $X = 0.2$, and the reaction/process has reached equilibrium).

Ans: In such case ($g/l + S$ reaction)/reactors, each particle can be considered to be segregated. There is no concentration distribution within a particle and particles do not mix with each other. The reactor can be considered to be macro or segregated reactor. Therefore,

$$\bar{X} = \int_0^\infty X(t) J'(t) dt$$

$$\downarrow$$

$$\text{(segregated model)}$$

Ans:

$$J'(t) = \frac{1}{10} e^{-t/10}$$

(No data are provided for RTD. So, assume well-mixed gas and solids in the reactor. Also note that the experimental measurements of RTD for solid phase is difficult, if not impossible)

$$\bar{X} = \int_0^\infty (0.2 + 0.26e^{-kt}) \times \frac{1}{10} e^{-t/10} dt$$

$$= \frac{1}{10} \int_0^\infty (0.2e^{-t/10} + 0.26e^{-1.1t}) dt$$

$$= \frac{1}{10} \left[0.2 \times -10e^{-t/10} \Big|_0^\infty - \frac{0.26}{1.1} \times e^{-1.1t} \Big|_0^\infty \right]$$

$$= \frac{1}{10} \left[2 + \frac{0.26}{1.1} \right] = \underline{0.223} \quad (or\ 22.3\%)$$

2) $\frac{dX}{dt} = -k\,(X - 0.2)$:

$$batch\ kinetics$$

$$ln(X - 0.2)|_{X_0=0.46}^{X} = -kt|_0^t$$

$$ln\left(\frac{X-0.2}{0.46-0.2}\right) = -kt$$

$$X = 0.20 + 0.26e^{-kt}$$

(batch reactor)

$$\left(\begin{array}{l} Note: t = 0\ ,\ X = 0.46 \\ \qquad\quad t \to \infty,\ X \to 0.20 \end{array}\right)$$

If there is mixing, the fluid elements lose their identities; they cannot be tracked.

There are J & J' (RTDs) for a mixed reactor, but they cannot be used to determine conversion:

$$\bar{X} \neq \int X(t)J'(t)dt \quad \left(\neq \sum X(t)J'(t)\Delta t\right)$$

Example

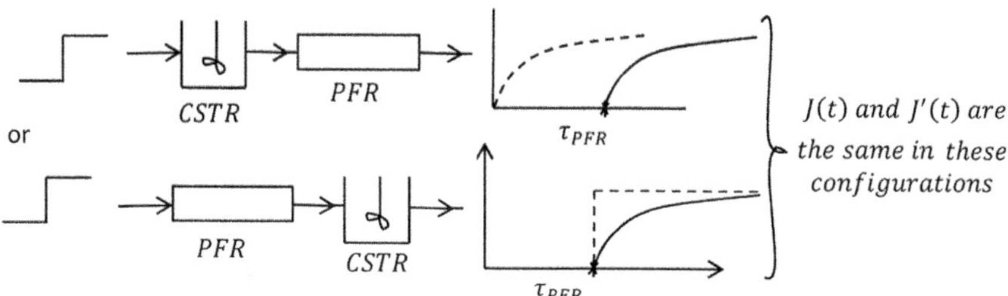

But, they do not yield the same conversion, and one cannot use

$$\bar{X} \neq \int X(t)J'(t)dt \quad \text{(first-order reaction is an exception)}$$

All it means is that 'RTD' is not good enough/sufficient to characterize the performance of the non-segregated or mixed reactors, or determine the conversion. One

© The Author(s) 2025
N. Verma, *Chemical Reaction Engineering*,
https://doi.org/10.1007/978-3-031-88691-1_20

also requires to quantify/determine the extent or rate of mixing, unlike the situation of macro-mixing in which case the segregated model uses only one parameter (RTD) to determine the conversion. Nonsegregated model has two parameters: RTD and the extent of mixing (rate of mixing or where and when mixing takes place, or the actual path of mixing). In other words, we need additional details of mixing. Average residence or mixing time alone is not good enough (1st order reaction is an exception).

Mechanistically, in mixed (non-segregated) reactors the average time a fluid element spends in the reactor is equally important as the actual path and speed of the fluid element to determine the conversion. Two non-segregated or mixed reactors may have the same RTD but different extent (early or late) of mixing and therefore, different conversions (1st order reaction is an exception).

Re-visit the two configurations consisting of an ideal CSTR and an ideal PFR. In one configuration, there is an early mixing followed by late mixing; in the other, late mixing is followed by early mixing. From the kinetic rate (order, $n > 0$) viewpoint, the first configuration represents the low-rate reactor followed by the high-rate reactor; the second configuration has the reverse situation. First-order reaction is an exception, discussed earlier. Therefore, an ideal CSTR reactor can still be treated as a segregated reactor for a first-order reaction and

$$\bar{X} = \int X(t)J'(t)dt$$
$$\downarrow$$
$$\text{batch reactor}$$

There are two commonly used non-segregated models

(a) Mixing Cells $\rightarrow$ two parameters: RTD and "n" (# of ideal CSTRs in series)
(b) Dispersion $\Rightarrow$ two parameters: RTD and "Pe" (or dispersion number)

In both models, the second parameter characterizes extent (early or late) of mixing or dispersion in the flow direction. In the extreme situation, $n = 1$ (ideal CSTR) and $n \rightarrow \infty$ (large) (ideal PFR), or $Pe\left(\frac{\bar{V}L_c}{D}\right) \rightarrow 0$ (ideal CSTR) and $Pe \rightarrow \infty$ (large) (ideal PFR). A non-ideal or real reactor assumes an intermediate value for n or Pe (in-between "infinite" and "zero" dispersion), i.e., neither an ideal CSTR nor an ideal PFR.

(a) **Mixing Cell Model:**

Measure:

(1) RTD of A (expt) (*Note*: $N = 1$ for CSTR and $\rightarrow \infty$ for PFR)
(2) RTD of "N" mix-reactors (analytical) $1 < n < N$

(3) Determine "N" by fitting the model with experimental data of the same RTD.

(4) Qualitatively fit the data (an exact or statistical fitting is not required).

Species balance over "nth" reactor

$$\begin{array}{c} C_o \quad \underline{\quad}\!\!\lceil\overline{\quad} \\ 0 \end{array}$$

$$v\,(C_{n-1} - C_n) = \left(\frac{V}{N}\right)\frac{dC_n}{dt}$$

i.e.,

$$t = 0^-, \quad C_n = 0 \text{ for all } n_s \text{ (no tracer)}$$

$$t = 0, \quad C_n = C_0(n = 0) \text{ (inlet to the 1st reactor)}$$

Use Laplace transformation

$$\mathcal{L}C_n(t) = \bar{C}(s)$$

$$v\left(\bar{C}_{n-1} - \bar{C}_n\right) = \frac{V}{N}s\bar{C}_n$$

(at the initial condition functional value is zero). Or

$$\left(1 + s\frac{\tau}{N}\right)\bar{C}_n - \bar{C}_{n-1} = 0 \quad \left(\tau = \frac{V}{v}\right)$$

$$\bar{C}_1 = \bar{C}_0/\left(1 + s\frac{\tau}{N}\right)$$

$$\bar{C}_2 = \bar{C}_1/\left(1 + s\frac{\tau}{N}\right) = \bar{C}_0/\left(1 + s\frac{\tau}{N}\right)^2$$

$$\bar{C}_n = \bar{C}_0/\left(1 + s\frac{\tau}{N}\right)^n$$

$$\bar{C}_n(s) = \bar{C}_0/\left(1 + s\frac{\tau}{N}\right)^n \Rightarrow C_n = \mathcal{L}^{-1}\left[\frac{C_0\left(\frac{N}{\tau}\right)^n}{s\left(s + \frac{N}{\tau}\right)^n}\right]$$

$$C_n = \mathcal{L}^{-1}\left[\frac{C_0\left(\frac{N}{\tau}\right)^n}{s\left(s + \frac{N}{\tau}\right)^n}\right]$$

$$J(t) \equiv \frac{C_n}{C_0} = 1 - e^{-\frac{Nt}{\tau}}\left[1 + \frac{Nt}{\tau} + \frac{1}{2!}\left(\frac{Nt}{\tau}\right)^2 + \cdots + \frac{1}{n-1!}\left(\frac{Nt}{\tau}\right)^{n-1}\right]$$

See the textbooks and draw qualitatively but accurately:

Input:

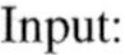

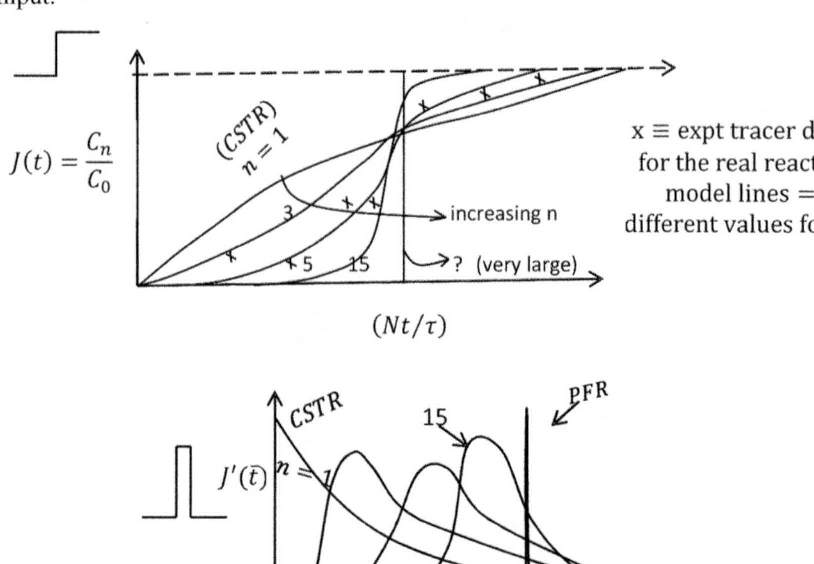

(Also note that $J'(t) = d(J(t))/dt \equiv$ same as analytically derived)

Model: Non-ideal (mixed) reactor of volume $V \equiv \,'N'$ ideal CSTR of volume V/N Once "N" is known, determine "X" for "N" ideal CSTR reactors (which have been discussed at the UG level).

Species balance (SS):

$$(C_{n-1} - C_n)v = \frac{V}{N}(-v_A r)$$
$$\downarrow$$
reaction rate

First-order reaction,

$$C_{n-1} - C_n = k\frac{\tau}{N}C_n$$

$$\left[\left(1 + \frac{k\tau}{N}\right)E - 1\right]C_n = 0 \ \text{(E is the shift operator)}$$

$$C_n = C_o \left(\frac{1}{1 + \frac{k\tau}{N}} \right)^n$$

$$\frac{C_o}{C_n} = \left(1 + \frac{k\tau}{N} \right)^n ; \quad C_n = C_o \left(1 - X_A \right)$$

For the 2nd or other order of reaction $\rightarrow$, use numerical technique.

(b) **Dispersion Model**

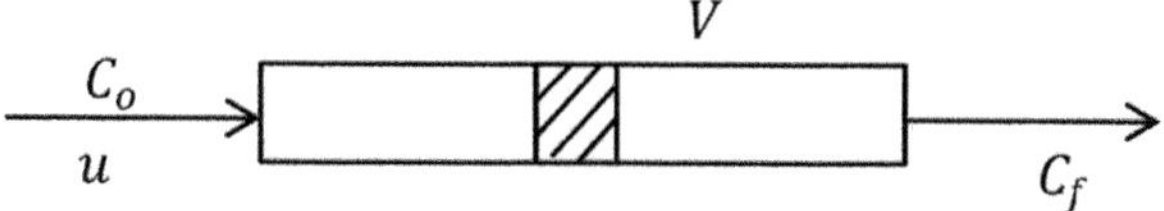

Non-ideality arises because of dispersion, implying local mixing along the flow direction, early or late. At high Reynolds numbers, there are turbulent flows and eddies. Also, there is a circulation of fluids in voids between packed bed materials, even at low Reynolds number.

The governing equation or model equation for dispersion is similar to that of molecular diffusion:

$$J = -D_m \nabla C : \text{molecular diffusion (Fick's 1st law)}$$

$$J = -D_L \nabla C : \text{dispersion}$$

D_L is axial dispersion coefficient; no radial dispersion:

$$\left(\frac{\partial}{\partial r} = 0 \right)$$

D_L is the model parameter.

How to determine D_L?

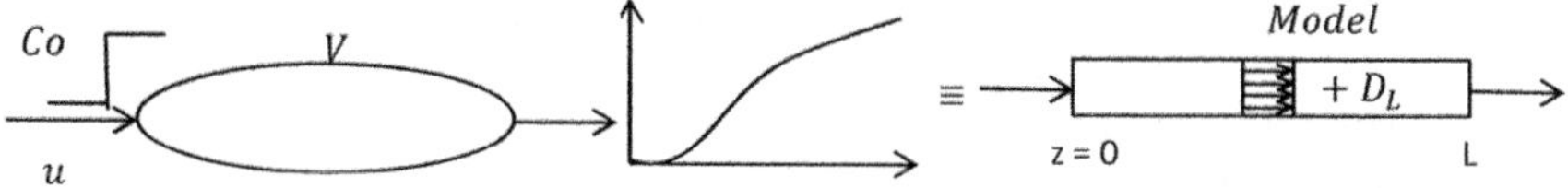

Note: $D_L = 0$ (no dispersion or mixing) in an ideal PFR.

Species balance for "tracer" (no reaction, transient experiment)

$$\frac{\partial C}{\partial t} = D_L \frac{\partial^2 C}{\partial z^2} - u \frac{\partial C}{\partial z}$$

$t = 0^-$

$$C = 0 \ (z \geq 0) \quad \text{(pure liquid)}$$

$t = 0$

$$z = 0, \quad uC_o = \left(uC - D_L \frac{\partial C}{\partial z} \right)\Bigg|_{z=0^+}$$

$$= L, \quad \frac{\partial C}{\partial z} = 0 \quad \text{(long tube approximation)}$$

(*Note:* $D_L \equiv$ dispersion coefficient of tracer)

Non-dimensionalize to get $J(t)$ and $J'(t)$ in terms of Pe

$$\frac{D_L}{uL} \frac{\partial^2 \phi}{\partial \eta^2} - \frac{\partial \phi}{\partial \eta} = \frac{\partial \phi}{\partial \bar{t}}$$

Recall:

$$Pe(axial\ peclet\ \#) \;=\; \frac{uL}{D_L} \quad \left(\frac{\text{convection}}{\text{dispersion}}\right)$$

large value implies small dispersion $\longrightarrow$ small value implies large dispersion

(PFR)   (CSTR)

(In general, $Pe > 10: PFR; \quad < 1 : CSTR$))

Solve to get C/C_o or $\phi(t)$ or $J(t)$ numerically:

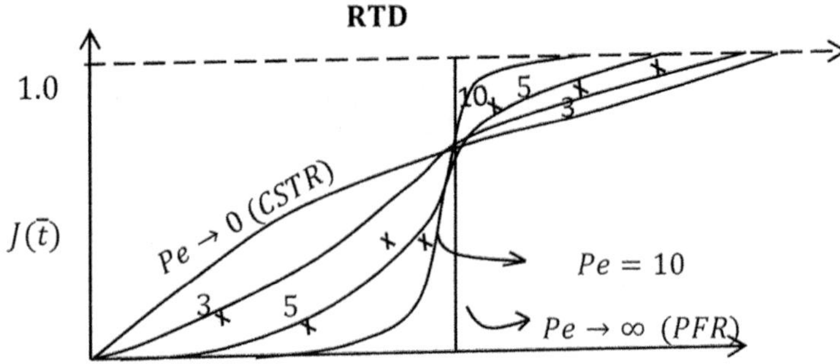

Note: Determine Pe by fitting the RTD data with the model predictions. For small D_L, or large Pe (not very large), analytical solutions are available.

$$J(\bar{t}) = \frac{1}{2}\left[1 - \mathrm{erf}\left(\frac{1 - \bar{t}}{2\sqrt{\bar{t}}} \right) \sqrt{Pe} \right]$$

$$\left(\mathrm{erf}(x) \left(= \frac{2}{\sqrt{\pi}} \int_0^x e^{-x^2} dx \right)\right)$$

$$J'(\bar{t}) = \frac{1}{2}\sqrt{\frac{Pe}{\pi}} \exp\left(-\frac{Pe}{4}(1 - \bar{t})^2 \right)$$

If you do a dose-experiment:

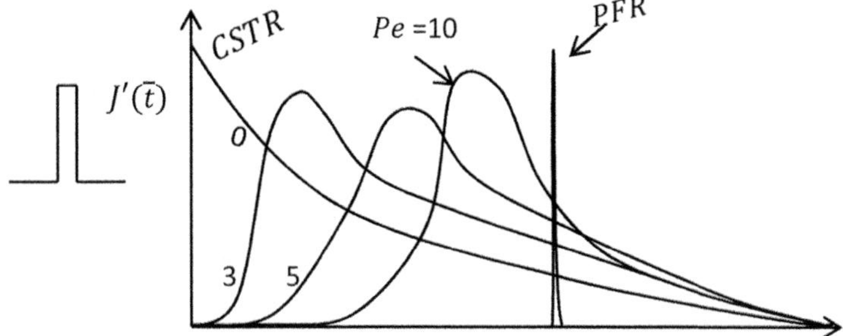

Note that in both cases (J or J'), the responses will be the same as before, *i.e.*, in mixed tanks in series model.

Once Pe or D_L is known, determine $\bar{X}$

$$\frac{\partial C}{\partial t} + u\frac{\partial C}{\partial z} = D_L\frac{\partial^2 C}{\partial z^2} + (-v_A r)$$

$$D_L\frac{d^2 C}{dz^2} - u\frac{dC}{dz} + (-v_A r) = 0$$

BC

$$z = 0 \quad uC_o = \left(uC - D_L\frac{dC}{dz}\right)\Bigg|_{z=0^+}$$

($C = C_o$ if no gradient (mixing) on $z = 0^+$ side)

A general interfacial boundary condition:

$$\left(uC - D_L\frac{dC}{dz}\right)\Bigg|_{L^-} = \left(uC - D_L\frac{dC}{dz}\right)\Bigg|_{L^+}$$

At the exit, continuity in concentration:

$$C_{L^-}=C_{L^+} \implies \frac{dC}{dz}\Bigg|_{L^-} = \frac{dC}{dz}\Bigg|_{L^+} = 0 \text{ (gradient is flat)}$$

For first-order reaction,

$$\frac{C_f}{C_o} = \frac{4a\exp(Pe/2)}{(1+a)^2\exp\left(\frac{aPe}{2}\right) - (1-a)^2\exp\left(-\frac{aPe}{2}\right)}$$

where

$$a = \sqrt{1 + \frac{4k\tau}{Pe}}$$

For the other reaction orders, numerical technique is required to solve the equation.

Summary

$$\text{non} - \text{ideal reactor}$$

(1) Find $J(t)$ for D_L or Pe.

(2) Determine

$$X\left(1 - \frac{C}{C_0}\right)$$

from the performance equation.

Note: Numerous empirical equations are available for packed bed reactors to determine $D_L = D_L(Re, Sc)$, or under turbulent flow conditions, $Pe_{,L} = Pe_{,L}(Re, Sc)$.

RTD study at a glance: It is an approach or a model to characterize or understand extent of mixing in a non-ideal continuous flow reactor, non-ideality arising out of either less than fully segregation or completely non-segregation in flow direction. Thus, there are two model ideal reactors as far as the effect of mixing on reaction conversion is concerned: an ideal CSTR or an ideal PFR; all real reactors stand in between. Therefore, "ideal reactors" imply that different fluid elements or pockets or zones are either completely mixed, non-segregated, dispersed, micromixed, or instantaneously mixed like in a well-mixed CSTR, or completely un-mixed, segregated, zero-dispersed in flow direction, macro-mixed like in a PFR without diffusion and dispersion in flow direction. RTD measurements (either step input or pulse input) can be used to characterize or describe the extent or state of mixing (early or late) in a real (or non-ideal) reactor, with an objective of predicting or calculating reaction conversion at the exit of the reactor.

If you have enough evidence (experimental or intuition or experience) to judge that a real reactor under consideration is segregated like in an ideal PFR or laminar tubular flow reactor, or fast fluidized gas/liquid-solid reactors or (immiscible) liquid-liquid membrane or extraction systems, the RTD data will suffice to calculate the conversion, and the performance or design or model equation of such non-ideal reactors is based on the RTD data (one parameter only) and the batch kinetics. However, if there is some element or extent of mixing, the RTD measurements alone will not suffice to describe the extent of mixing in such non-ideal reactors, and additional information is required: "where do they (different pockets) mix and at what rate do they mix". Thus, there are two popular models: "tank-in-series-model" with the number of ideal CSTRs as the model parameter and "Dispersion model" with peclet or dispersion number as the model parameter. Once the RTD and the model parameter are known, reaction conversion can be calculated for such non-ideal (non-segregated) reactors from the first principle (species balance equation). First-order reaction is an exception, and the RTD data (with batch kinetics) alone are enough to calculate or predict the conversion in non-segregated reactors.

Fluid-Particles Reactions (Non-catalytic)

Heterogeneous reaction $(2 - \phi)$:

$$(-r_A) = \underbrace{kC_{A,S}}_{} \neq \underbrace{kC_{A,b}}_{}$$

at the solid surface $\quad$ bulk g/l phase

1. Size of unreacted solid decreases:

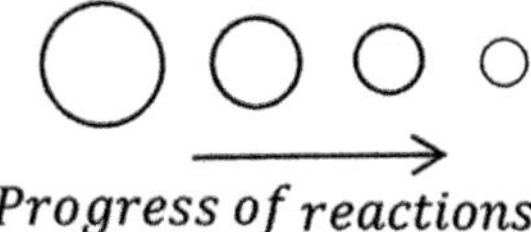

Progress of reactions

2. Ash-layer formation over unreacted solid.

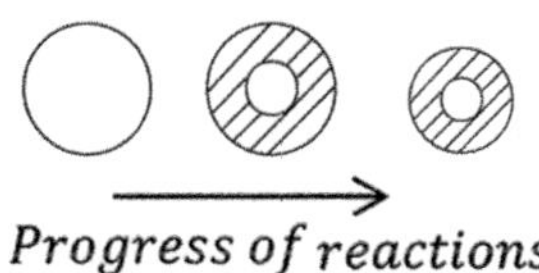

Progress of reactions

Examples:

$$(a) \quad 2ZnS + 3O_2(g) \longrightarrow 2ZnO(s) + 2SO_2(g)$$

$$(b) \quad 4FeS_2(s) + 11O_2(g) \longrightarrow 8SO_2 + 2Fe_2O_3(s)$$

N. Verma, *Chemical Reaction Engineering*,
https://doi.org/10.1007/978-3-031-88691-1_21

(c) $Fe_3O_4(s) + 4H_2(g) \longrightarrow 3Fe(s) + 4H_2O(g)$

(d) $CaC_2(s) + N_2(g) \longrightarrow CaCN_2 + C(amorphous)$

$\begin{cases} (e) & C(s) + H_2O(g) \longrightarrow CO(g) + H_2(g) \\ (f) & C(s) + 2H_2O(g) \longrightarrow CO_2(g) + 2H_2(g) \end{cases}$ products are gases

1. Progressive-conversion model
2. Unreacted-Core Model (Shrinking "Core" Model), where "core" is non-porous, and shell is product (ash).

Shrinking-Core Model

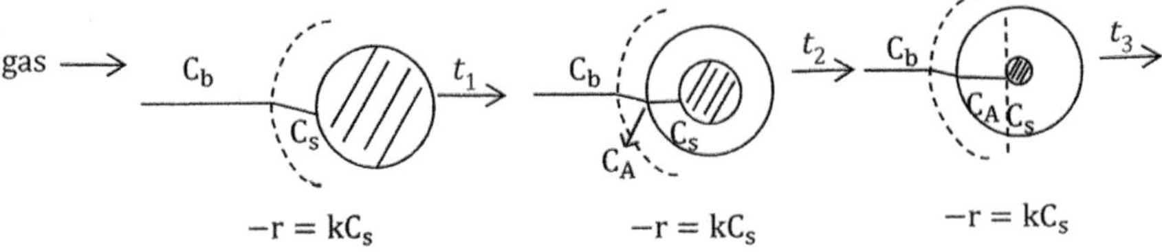

(1) Gas diffuses from bulk to the surface (resistance)

$$J = k_m \,(C_b - C_s) \quad k_m = f(Re, Sc)$$

(2) Gas has to diffuse through the ash-layer,

$$J = -D_s \frac{\partial C_A}{\partial r}$$

(3) Gas reacts at the surface,

$$-r = kC_s \quad (C_b > C_s) \quad \text{(irreversible)}$$

where, C_s gas-phase concentration at or near the surface.

$$\underset{\text{(gas)}}{A} + \underset{\text{(solid)}}{bB} \xrightarrow{kr} \text{Products}$$

Case (1) Diffusion through gas film controls

(2) Diffusion through ash layer controls

(3) Reaction rate controls

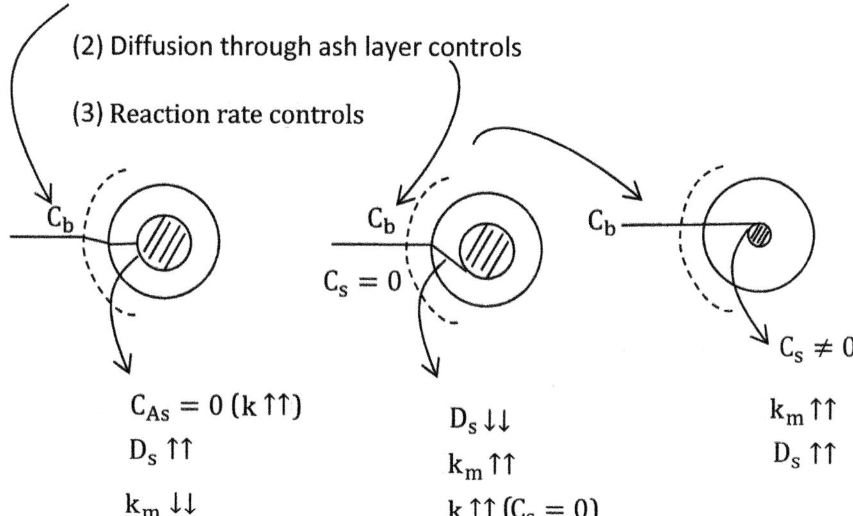

Case 1:

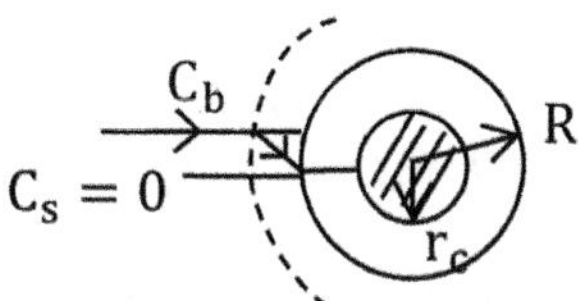

Define X_B (based on solid):

$$= \frac{\left(\frac{4}{3}\pi R^3 - \frac{4}{3}\pi r_c^3\right)}{\left(\frac{4}{3}\pi R^3\right)} = 1 - \left(\frac{r_c}{R}\right)^3$$

$$X_B(t) \rightarrow r_c(t)$$

Species Balance: (Flux × Area balance)

$$J_{Flux}\,\text{of}\,A(gas) = k_m\,(C_b - C_s)$$

$$\frac{moles}{s - m^2} = \underbrace{k_m}_{m/s}\,\underbrace{C_b}_{moles/m^3}\quad (C_s = 0)$$

Rate of A reaching the surface (N_A):

$$k_m C_b 4\pi R^2 \left(\frac{moles}{s}\right)$$

Rate of B reacting (N_B):

$$\frac{d}{dt}\left(\frac{4}{3}\pi r_c^3\right) \times \rho_B \left(\frac{gm}{cc}\right) \times \frac{1}{M_B}\left(\frac{moles}{s}\right)$$

$$b\left(4\pi R^2\right)k_m C_{Ag} = -\frac{d}{dt}\left(\frac{4}{3}\pi r_c^3\right) \times \frac{\rho_B}{M_B} = -\frac{4\pi\rho_B}{M_B}r_c^2\frac{dr_c}{dt}$$

$$bN_A = N_B \text{ moles/s}$$

$$b\left(4\pi R^2\right)k_m C_b = -\frac{d}{dt}\left(\frac{4}{3}\pi r_c^3\right) \times \frac{\rho_B}{M_B} = -\frac{4\pi\rho_B}{M_B}r_c^2\frac{dr_c}{dt}$$

$$bk_m C_b t = \frac{1}{3}\frac{\rho_B}{M_B R^2}\left(R^3 - r_c^3\right)$$

$$t = \frac{\rho_B R}{3M_B k_m C_b b}\left(1 - \left(\frac{r_c}{R}\right)^3\right)$$

$$X_B = t\left(\frac{3M_B k_m C_b b}{\rho_B R}\right)$$

Define τ to completely react (core vanishes), $r_c \to 0$

$$\tau = \frac{\rho_B R}{3M_B k_m C_b b} = \frac{\rho_B R}{3M_B k_m b C_b}$$

$$\boxed{X_B = \frac{t}{\tau}}$$

$$\boxed{\frac{t}{\tau} = 1 - \left(\frac{r_c}{R}\right)^3 = X_B}$$

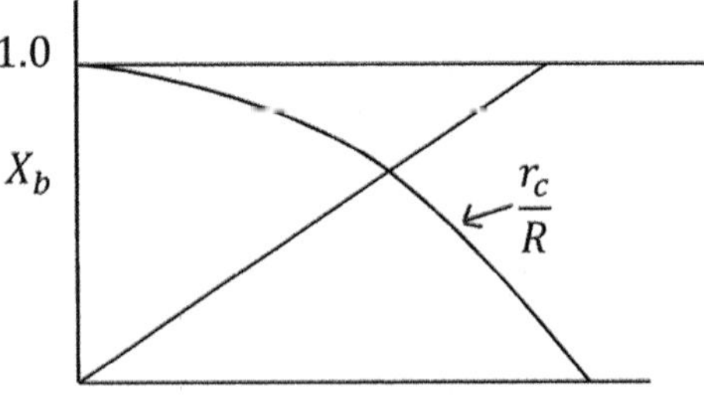

Case 2: Ash layer diffusion-control ($k_m \uparrow, k_r \uparrow, D_s \downarrow$)

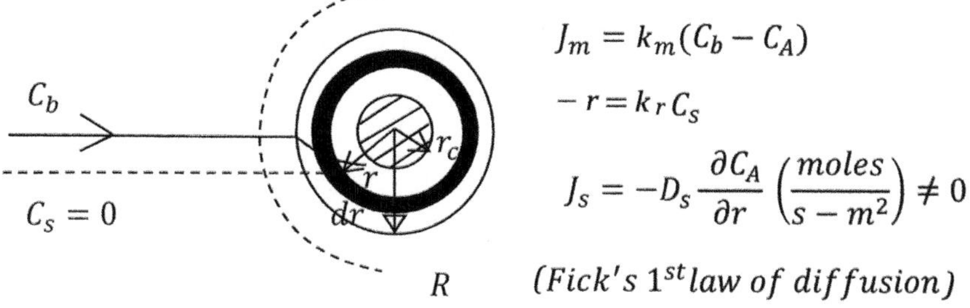

$$\left. \begin{array}{c} r_c(t) \leq r \leq R \\ 0 \leq r_c \leq R \end{array} \right]$$

Note: $k_m \longrightarrow \infty$; $C_b \longrightarrow C_A$ and $k_r \longrightarrow \infty$; $C_s \longrightarrow 0$.

Rate:

$$\left\{ \begin{aligned} \frac{dN_A}{dt} &= (\text{ no of moles of } A)/s \qquad (1) \\ &= -D_s \frac{\partial C_A}{\partial r}\bigg|_{r=R} \times 4\pi R^2 \\ &= -D_s \frac{\partial C_A}{\partial r}\bigg|_{r=r} \times 4\pi r^2 \qquad (2) \\ &= -D_s \frac{\partial C_A}{\partial r}\bigg|_{r=r_c} \times 4\pi r_c^2 \\ &= \text{Const.} \end{aligned} \right.$$

$$\begin{aligned} \left(\frac{dN_B}{dt}\right) &= \frac{d}{dt}\left(\frac{4}{3}\pi r_c^3 \times \frac{\rho_B}{M_B}\right) \\ &= \frac{4\pi r_c^2 \rho_B}{M_B} \frac{dr_c}{dt} \qquad (3) \end{aligned}$$

PSS:

$$-\frac{D_S(C_b - 0)}{\left(\frac{dN_A}{dt}\right)} = \frac{1}{4\pi}\left(\frac{1}{R} - \frac{1}{r_c}\right)$$

$$-\frac{dN_A}{dt}\left(\frac{1}{r_c} - \frac{1}{R}\right) = 4\pi D_S C_b \qquad (4)$$

$$b N_A = N_B \quad (\text{Recall})$$

$$t = \frac{\rho_B R_p^2}{6bD_s C_b}\left[1 - 3\left(\frac{r_c}{R}\right)^2 + 2\left(\frac{r_c}{R}\right)^3\right]$$

or,

$$\frac{t}{\tau} = 1 - 3\left(\frac{r_c}{R}\right)^2 + 2\left(\frac{r_c}{R}\right)^3 \Rightarrow \frac{t}{\tau} = 1 - 3(1 - X_B)^{\frac{2}{3}} + 2(1 - X_B)$$

Case 3: Chemical reaction controls $k_m \uparrow, D_s \uparrow, k \downarrow$

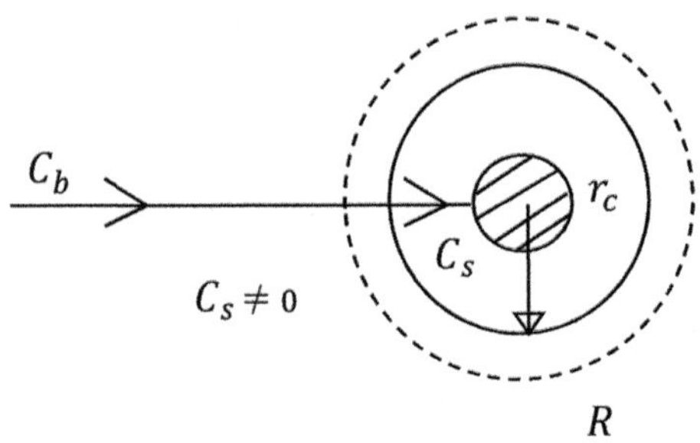

$$- r = k_r C_b \left(1^{st}\text{-order}\right)$$

$$J_m = k_m \left(C_b - C_s\right) = \text{ finite}$$

$$k_m \to \infty, \ C_b = C_A = C_s$$

$$J_S = -D_S \frac{\partial C_A}{\partial r} \ (r_c \leq r \leq R)$$

$$D_S \to \infty, \ \ \frac{\partial C_A}{\partial r} \to 0: \ \ \text{No gradient inside the ash layer}$$

Under SS,

$$(-r_A) = J_m \ (\text{moles/s-cm}^2)$$

Rate of B consumed

$$= \frac{d}{dt}\left(\frac{4}{3}\pi r_c^3 \times \frac{\rho_B}{M_B}\right) = \text{ finite}$$

$$= \frac{4\pi r_c^2 \rho_B}{M_B}\frac{dr_c}{dt}\left(\frac{\text{moles}}{s}\right)$$

Rate of A consumed $(C_b = C_A = C_s)$

$$-r_A = \left(k_r C_s\right) \times 4\pi r_c^2 \left(\frac{moles}{s-m^2}\right)$$

$$flux \qquad \frac{m}{s} \qquad \frac{moles}{m^3}$$

$$b4\pi r_c^2 k_r C_b = \left(\frac{4\pi r_c^2 \rho_B}{M_B}\frac{dr_c}{dt}\right) : \text{ stoichimetry}$$

$$k_r C_b t = \frac{a\rho_B}{M_B} R_p \left(1 - \frac{r_c}{R}\right)$$

$$t = \frac{\rho_B R_p}{k_r C_b M_B b}\left(1 - \frac{r_c}{R}\right) \Rightarrow t = \tau\left(1 - \frac{r_c}{R}\right)$$

or

$$\boxed{\frac{t}{\tau} - 1 - (1 - X)^{\frac{1}{3}}}$$

Example 1

Consider the non-catalytic gas-solid phase (first-order rate) reaction

$$A\,(g) + bB\,(s) \rightarrow \text{product(s)}$$

Under the present conditions, the ash-layer diffusion resistance is negligible, but both the gas-film resistance and surface reaction are important. Determine the time of complete conversion of the solid in terms of R, r_c, k_m, k, ρ_B, M_B and C_b; these parameters have usual meaning. Also, determine $X_B\,(t)$.

Ans:

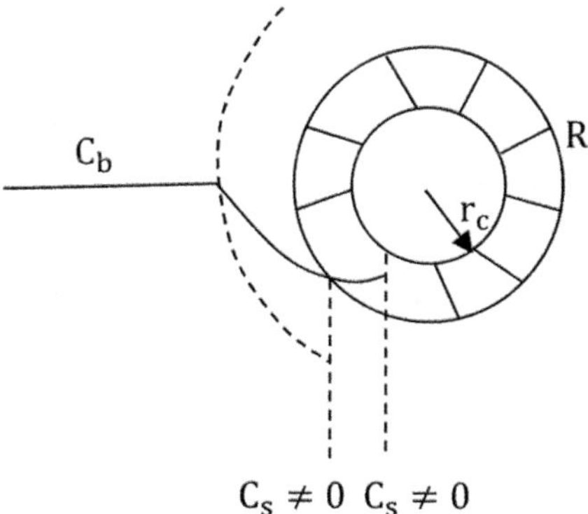

$$A\,(g) + bB\,(s) \rightarrow \text{Products}$$

$$- r_A = k C_A \left(\frac{\text{moles}}{\text{s-m}^2} \right)$$

$$\begin{cases} r_c \equiv \text{ radius of core (unreacted material)} \\ R = \text{Outer radius of the solid (particle)} \\ \text{Ash layer: } r_c \leq r \leq R \end{cases}$$

$\Rightarrow$ Rate of moles of A arriving at the solid surface (moles/s);

$$\underbrace{4\pi R^2 k_m}_{} (C_b - C_s) = \underbrace{4\pi r_c^2 k}_{} C_s$$

$$\text{m}^2\text{m/s moles/m}^3 \qquad \text{m}^2\text{m/s moles/m}^3$$

(rate of moles of A reacting at the core surface)

$$k_m \equiv \text{ film-diffusion (mass-transfer) coefficient (m/s)}$$

$$k = \text{rate constant} \left(\frac{\text{m}}{\text{s}} \right)$$

$$C_b = \text{bulk phase gas } (A) \text{ concentration} \left(\frac{\text{moles}}{\text{m}^3} \right)$$

$$C_s = \text{gas-phase gas } (A) \text{ concentration at } r = r_c \left(\frac{\text{moles}}{\text{m}^3} \right)$$

$$\Longrightarrow C_s = \left(\frac{R^2 k_m}{R^2 k_m + r_c^2 k} \right) C_b = \left(\frac{k_m}{k_m + (r_c/R)^2 k} \right) C_b$$

Note: the ash-layer $(r_c - R)$ diffusion resistance is negligible.

$\Longrightarrow$ No concentration-gradient in ash-layer, ∇C_A across $(r_c - R) \cong 0$.
Equate the rate of change in the particle size with the rate of reaction:

$$- \frac{d}{dt} \underbrace{\left(\frac{4}{3}\pi r_c^3 \frac{\rho_B}{M_B} \right)}_{} = b \left(4\pi r_c^2 k \right) \left(\frac{k_m}{k_m + \left(\frac{r_c}{R}\right)^2 k} \right) C_B$$

$$\downarrow$$

$$\text{Volume of core} \times \left(\begin{array}{l} \rho_B = \text{Bulk density of solid (g/cm}^3) \\ M_B = \text{Molecular weight of solid} \end{array} \right)$$

Moles of B reacting with gas:

$$(A) = \underbrace{b \left(4\pi r_c^2 \right) k C_s}_{}$$

$$\text{m}^2\text{m/s moles/m}^3$$

where C_s is substituted in terms of C_b.

Or

$$\boxed{-\frac{dr_c}{dt} = \frac{\alpha}{\beta + r_c^2}}$$

$$\alpha = \left(\frac{bR^2 k_m M_B}{\rho_B}\right) C_B; \quad \beta = \left(\frac{k_m}{k}\right) R^2 \quad \text{(Check the unit)}$$

On integration,

$$-\int_R^{r_c} \left(\beta + r_c^2\right) dr_c = \int_0^t \alpha dt$$

$$\beta\left(R - r_c\right) + \left(\frac{R^3}{3} - \frac{r_c^3}{3}\right) = \alpha t$$

Definition: $t \to \tau$, $r_c \to 0$ (complete conversion). Therefore,

$$\tau = \frac{\left(\beta R + \frac{R^3}{3}\right)}{\alpha}$$

and

$$X_B(t) = 1 - \frac{r_c^3}{R^3}$$

Example 2

Consider the reaction ($T = 900°C$, $P = 1$ atm):

$$Fe_3 O_4\,(s) + 4H_2 O\,(g) \longrightarrow 3Fe\,(s) + 4H_2 O\,(g)$$

Neglecting the ash-layer diffusion resistance, calculate the time for the $Fe_3 O_4$ particle size (radius) to decrease from 5 mm to 1.25 mm. Concentration of $H_2 = 4\%\,(v/v)$, Bulk density of

$$F_3 O_4 = 5000\frac{\text{kg}}{\text{m}^3}, \quad M_{Fe_3 O_4} = 230\frac{\text{gmol}}{,} \quad k_m = 1\frac{\text{cm}}{\text{min}}$$

$$k = 1\,\frac{\text{cm}}{\text{min}}, \quad R_g = 0.082\,\frac{\text{L} - \text{atm}}{\text{K} - \text{mol}}$$

Ans: This is the similar prototype case as in the previous example. Mechanistically, ash-layer diffusion is negligible, but film (external)-mass-transfer or diffusion resistance and kinetic rate are important (significant).

$$-\frac{dr_c}{dt} = \frac{\alpha}{\beta + r_c^2}; \quad \alpha = \left(\frac{bR^2 k_m M_B}{\rho_B}\right) C_b; \quad \beta = \frac{k_m R^2}{k}$$

Here,

$$b = \frac{1}{4}; \quad R = 0.005 \text{ m}, \quad k_m = 0.01\frac{\text{m}}{\text{s}}$$

$$M_B \, (Fe_3O_4) = 230, \quad \rho_B = 5000 \,\frac{\text{kg}}{\text{m}^3};$$

$$C_b = \frac{p_b}{RT} = \frac{4}{100} \times \frac{1}{0.082 \times 10^{-3} \times (273 + 900)} = 0.416 \text{ moles/m}^3$$

$$\left.\begin{array}{l} \alpha = 1.195 \times 10^{-9} \\ \beta = 0.000025 \end{array}\right\}$$

Integrate:

$$\beta \, (R - r_c) + \left(\frac{R^3}{3} - \frac{r_c^3}{3}\right) = \alpha t$$

$$R = 0.005 \text{ m}, \quad r_c = 0.00125 \text{ m}$$

Substitute $t = 43$ min.

Example 3

Uniform-sized spherical particles VO_3 are reduced to VO_2 in a uniform environment with the following results:

$t(h)$	0.180	0.347	0.453	0.567	0.733
X_B	0.450	0.68	0.80	0.95	0.98

If the reaction follows SCM, find the controlling mechanism and a rate equation to represent this reduction (Levenspiel)

Ans: Recall the experiment for three cases:

(a) Gas film or external (mass-transfer) diffusion controls:

$$\frac{t}{\tau} = X_b \quad (k \uparrow\uparrow, \; D_{ash} \uparrow\uparrow, \; k_m \downarrow\downarrow)$$

(b) Reaction controls ($D_{ash} \downarrow\downarrow, k_m \uparrow\uparrow, \; k \downarrow\downarrow$)

$$\frac{t}{\tau} = 1 - (1 - X_b)^{\frac{1}{3}}$$

(c) Ash-film (layer) controls ($D_{ash} \downarrow\downarrow, k_m \uparrow\uparrow, \; k \uparrow\uparrow$)

$$\frac{t}{\tau} = 1 - 3(1 - X_b)^{\frac{2}{3}} + 2(1 - X_b)$$

Also, the mechanism which controls will have the rate expression containing the rate parameter/constant; the other parameters will drop out. Thus,

$$\tau_a \propto \frac{1}{k_m}; \quad \tau_b \propto \frac{1}{k}; \quad \tau_c \propto \frac{1}{D_e}$$

(τ_s will also have the dependence on particle size). For such problem, there is no straight solution: Trial and Check.

$t(h)$	X_B	$1 - (1 - X_b)^{1/3}$	$1 - 3(1 - X_b)^{2/3} + 2(1 - X_b)$
0.18	0.45	0.1807	0.0861
⋮	⋮	⋮	⋮
0.73	0.98	0.7286	0.819

Plot $f(X_i)$ vs t to check linearity

$$\left(y = mx; \quad m = \frac{1}{\tau} \right)$$

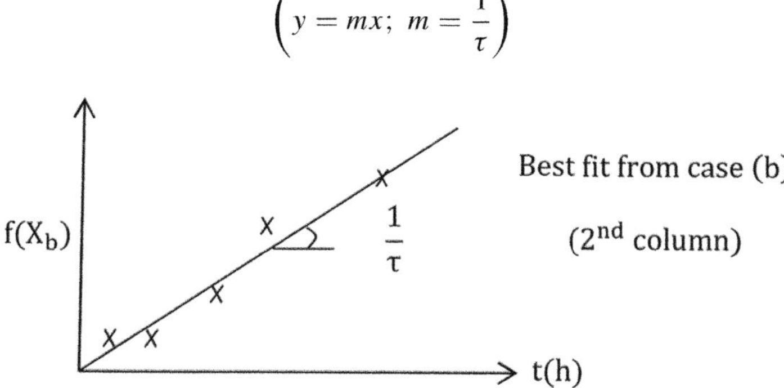

Reaction controls:

$$\tau = \left(\frac{\rho_b R}{bk C_b} \right), \quad \left(\text{or } \tau \propto \frac{R}{k} \right)$$

- What are catalysts? These are physical entities; show chemical/physical activities of deactivation, degradation, change in morphology, and have lifetime.
- What do they do? Enhance (+ve)/retard (−ve) rate of reaction by participating at the molecular or electronic levels, yet preserving/restoring their initial activities.
- How do they do? Direct alternate path (route/step) of reaction that has a lower activation energy barrier:

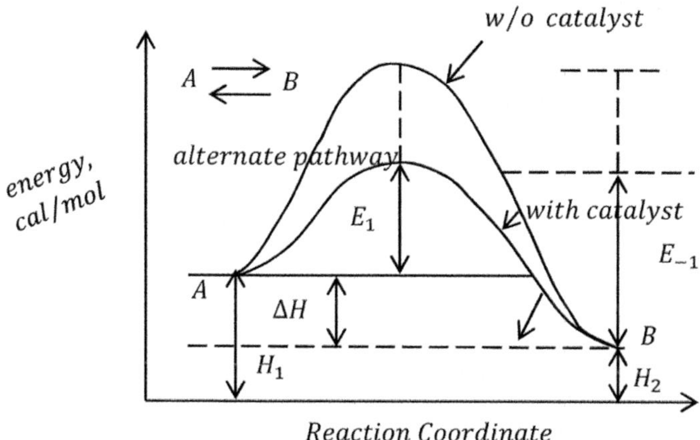

© The Author(s) 2025

N. Verma, *Chemical Reaction Engineering*,

https://doi.org/10.1007/978-3-031-88691-1_22

1. The barrier has come down using catalyst
2. There are two barriers: forward & reverse reactions (E_1 & E_{-1})
3. H_1 & H_2 or ΔH is the same with/without catalyst

- Catalyst enhances reaction rate or kinetics (k) of an elementary reaction without affecting thermodynamic properties including heat of reaction & equilibrium constant (K).
- No reaction is truly irreversible.
- ΔH can be $+ve$ or $-ve$ (endothermic or exothermic)

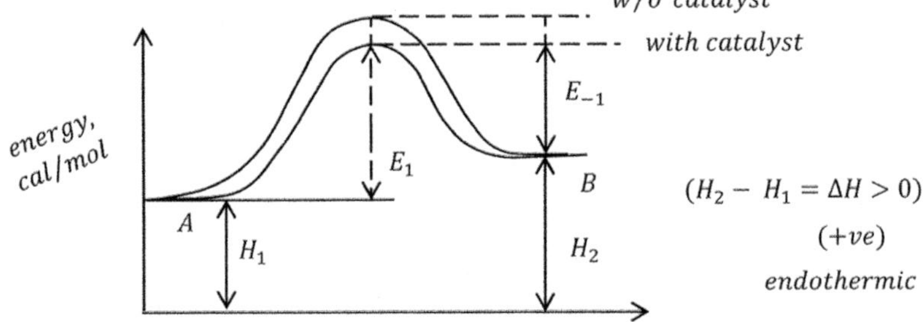

Note: in both cases

$$| (E_1 - E_{-1})| = | (H_2 - H_1) |$$

with/without catalyst.
- What are the features of a catalyst? Large S/V, $+ve$ or $-ve$, selectivity $A \rightarrow R$; may or may not require support; may require promoters (K, Na); there are inhibitors (poisons) (Cs, S, Pb) as well.
- Classification (types). For example

Metal catalyst:

$$C_2H_5OH \xrightarrow{Pt} C_2H_4O + H_2 \quad \text{(dehydrogenation)}$$

Acid catalyst:

$$\text{(electron acceptor/proton donor)} \xrightarrow{Si/Al} C_2H_4 + H_2O \text{ (dehydration)}$$

Bi-functional catalyst

Zeolites (catalytic reforming)

Cage-like structure

- History of the catalysts?

(1) First catalytic reaction (1875)

$$SO_2 + \frac{1}{2}O_2 \xrightarrow{Pt} SO_3 \quad \text{(Contact process)}$$

$(V_2O_5 + K_2SO_4)$ supported on Al_2O_3/SiO_2

(2)

$$NH_3 + O_2 \xrightarrow[900^oC]{Pt/10\%Rh} NO, NO_2 \quad (1903)$$

$$\searrow natural\ impurity\ in\ Pt$$

$$2NO + O_2 \rightarrow 2NO_2 \quad (HNO_3 \text{ production })$$

(3)

$$N_2 + 3H_2 \xrightarrow[500-600^\circ C]{Fe} 2NH_3 \quad \text{(Haber-Bosch Reaction)}$$

(4)

$$CO + H_2 \xrightarrow{ZnO/Cr_2O_3 \text{ (promoter)}} CH_3OH \quad (1923)$$

(5)

$$CO + H_2 \xrightarrow[Fe]{Ni} \underbrace{C_XH_y + H_2O}_{\text{(sythetic fuels)}} \quad \text{(Fischer-Tropseh 1930)}$$

(6) Catalytic cracking (1937) in petroleum industry

Al_2O_3/SiO_2: Refractory materials, Zeolites (molecular sieves)

(7) Reforming: Butane $\rightarrow$ Isobutane (polymer, plastics) (1950)

$$Mo_2O_3/Al_2O_3$$
$$Pt/Al_2O_3$$

(8) Hydrocracking (1960) $\Rightarrow$ Shell (catalyst mixed with H_2 to prevent formation or deposition on catalysts)

(9) Catalytic convertors $(60 - 70)$ in automobile exhaust

- Very complicated design (unsteady-state conditions of speed, temperature, species, no control), CO, C_XH_Y, NO (pollutants)

- $$CO + \frac{1}{2}O_2 \rightarrow CO_2$$

$$C_X H_y + \frac{1}{2}O_2 \rightarrow CO_2 + H_2O \quad Pt/Rh \text{ on } \underbrace{Al_2O_3/SiO_2}_{\text{Bauxite}}$$

$$(2NO + O_2 \rightarrow 2NO_2)$$

$$+ CO \rightarrow CO_2 + N_2)$$

O_2 disproportional reaction

- Modern vehicles have three-ways functional catalysts

Mechanism (Pathway)

$\left\{ \begin{array}{l} \text{all of these involve} \\ \text{several electron} \\ \text{transfer steps, complex} \\ \text{bonds formation} \end{array} \right.$ Electronic (sites are electronically active); geometrical theory (molecular dimensions/configuration)

Chemical theory (surface complex formation)
$\downarrow$
underlying solid surface should accomodate
$$(A \rightarrow A^*)$$

All these different theories and mechanism to help us drive a rate expression from the viewpoint of designing the reactor:

$$r = f(T, C)$$

Notes

1. Revisit:

$$r = \frac{1}{v_i} \frac{1}{S} \frac{dN_i}{dt}$$

S is capacity (A, W, V).

2. Collision on molecular level.

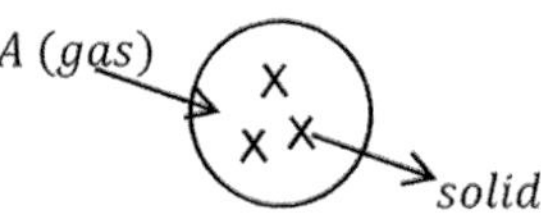

(It is obvious that transport is not a right step)

3. Unfortunately, there are transport steps from gas to solid for reactant before reaction, and from solid to gas for products after reaction:

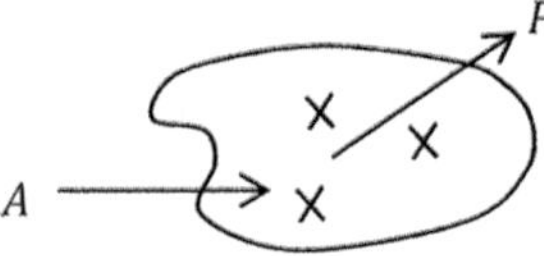

(Intitutively, the transport of A or P retards the overall rate of reaction)

4. Reaction may be isothermal or temp gradient may exist; can be exothermic or endothermic.

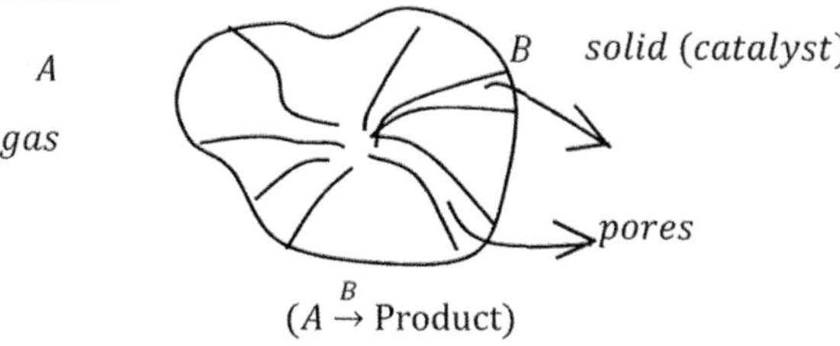

$$(A \xrightarrow{B} \text{Product})$$

transport steps
$\left\{\begin{array}{l}\\ \\ \\ \\ \\ \\ \\ \\ \\ \end{array}\right.$

   a. Bulk transport of 'A' towards solids (convection, diffusion)

$$(V . \nabla C \quad D\nabla^2 C)$$

   b.   ↓ Interphase transport (film transport)

                k_m *B (external surface)*

   (**A** encounters diffusion resistances) $\underline{k_m \Delta C \ (-D_m \nabla C)}$

   c. Intraphase transport $\left(\text{within pores}, \underline{\text{pore diffusion}}\right)$

     (reaction may take place parallel) $D_{pore} \nabla^2 C$

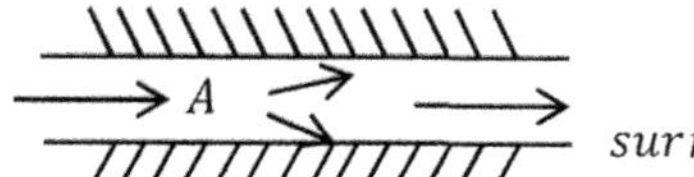

   d. Surface reaction (P is produced)

$$\left.\begin{array}{l}e. \\ f. \\ g.\end{array}\right\} \equiv \left.\begin{array}{l}c \\ b \\ a\end{array}\right\} \text{ for product P (transport steps)}$$

<u>surface reaction</u> (preceded and followed) by adsorption/desorption

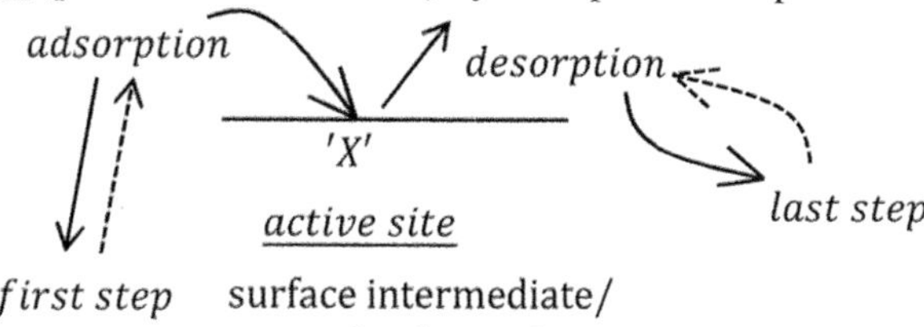

first step   surface intermediate/
               complex formation
               (several possibilities)

Adsorption (first step)/Desorption (last step):

(1) Langmuir model (1918)

Assumptions:

- Maximum one layer or monolayer
- Uniform surface activity
- No interaction between adsorbing molecules
- Constant activation energy and heat of adsorption, with respect to surface coverage, or $\neq f(\theta)$

$$r_A = \frac{sp}{(2\pi mKT)^{1/2}} \text{ (adsorption rate)}$$

$$s = \alpha(1 - \theta)e^{-E_a/RT}$$

$$0 \leq \theta \leq {}^V\!/_{V_m} \leq 1$$

p = Partial pressure of adsorbate

$$m = \left({}^{Mw}\!/_N\right)$$

K = Boltzman constant

α = sticking Coefficient

θ = fractional surface coverage

E_a = activation energy of adsorption

V = present amount,

V_m = full amount to cover the surface

vacant sites

$$r_a = k_a p(1 - \theta)$$

$$(k_a = k_{a_o} e^{-E_a/RT})$$

$$r_d = k_d \theta \text{(desorption rate)}$$

$\Rightarrow$ At equilibrium

$$r_a = r_d$$

$$k_a p(1 - \theta) = k_d(\theta)$$

$$\theta = \frac{k_a/k_d \, p}{1 + k_a/k_d \, p}$$

$$\boxed{\theta = \frac{Kp}{1 + Kp}}$$

$$(k_d = k_{do} e^{-E_d/RT})$$

N. Verma, *Chemical Reaction Engineering*,
https://doi.org/10.1007/978-3-031-88691-1_23

The model is good for chemisorption (one layer or monolayer)

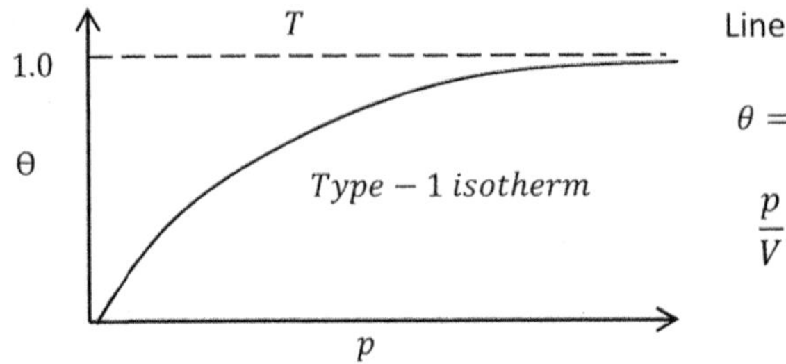

Linearized form:

$$\theta = {}^{V}\!/_{V_m} = \frac{Kp}{1 + Kp}$$

$$\frac{p}{V} = \frac{1}{KV_m} + \frac{1}{V_m}p$$

No interactions between adjacent molecules:

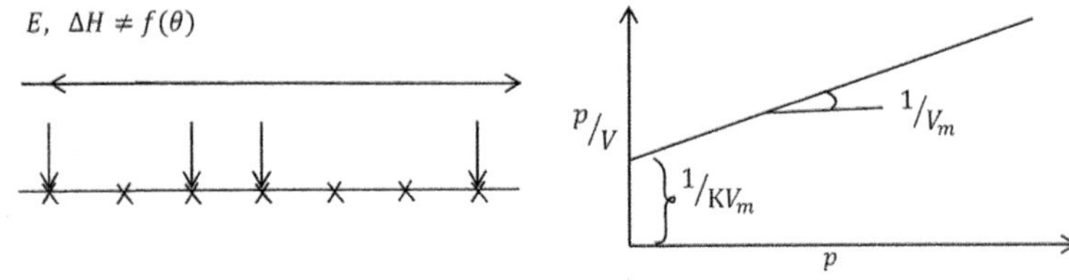

(2) Brunauer, Emmett and Teller (1938)

BET isotherm: Assumptions

- Multilayer adsorption
- Energy of first layer $\neq$ that of second layer; second and the other layers have the same energy level.
- We do not assume that first layer is saturated and then second layer is formed
- Finite probability for sitting of one molecule on the vacant or occupied sites.

Isotherm:

$$\theta = \frac{V}{V_m} = \frac{cP}{(p_o - p)\left[1 + (c - 1)\frac{p}{p_o}\right]}$$

$V =$ Volume adsorbed

$V_m =$ Volume adsorbed for one layer

$c =$ parameter of model

$$= \frac{\text{(energy of the first layer)}}{\text{(energy of the other layers)}} = 1 \text{ if the layers have the same energy.}$$

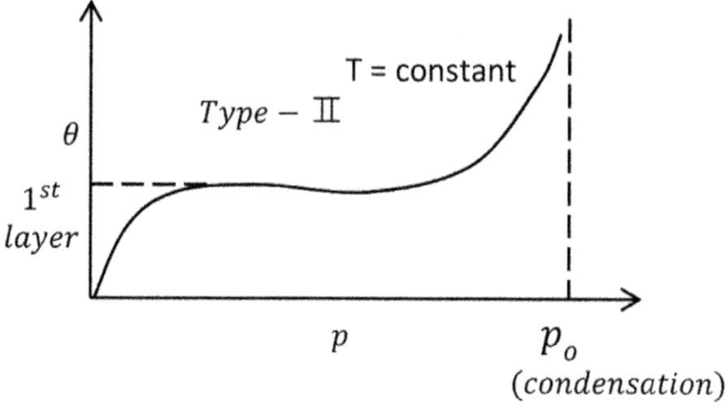

If $c \gg 1$ (very energetic surface) and $p \ll p_o$

$$\theta = \frac{c\,(p/p_o)}{1 + c\,(p/p_o)}$$

$$= \frac{Kp}{1 + Kp} \quad \text{(identical to Langmuir model)}$$

Linearize:

$$\frac{p}{V\,(p_o - p)} = \frac{1}{V_m C} + \frac{(c-1)p}{c\,V_m\,p_o}$$

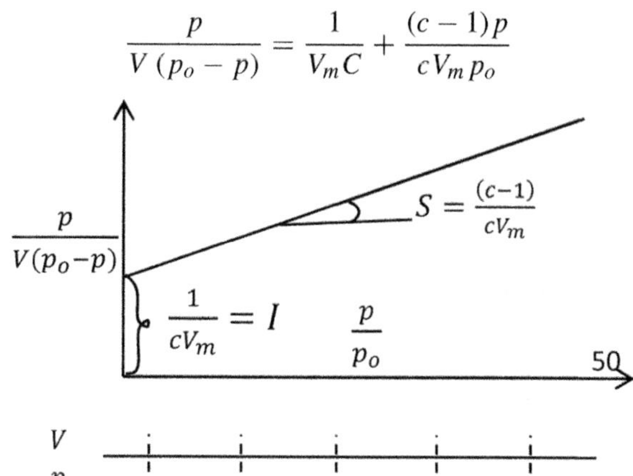

Data:

$$\begin{array}{c} V \\ p \end{array}$$

(3) Temkin Model:

$$r_a = k_a p \quad : \quad \begin{cases} r_d = k_d\theta = \left(k_{do}e^{-E_d/RT}\right)\theta \\ \left(E_d = E_{d_o} - b\theta;\ \theta \uparrow\ E_d \downarrow\right) \\ \exp\left(b'\theta\right)\theta \end{cases}$$

At equilibrium,

$$k_a p = A \exp\left(b'\theta\right)\theta$$

$$\theta \propto \ln(p)$$

Similarly,

$$\Delta H\,(E_d - E_a) \propto f(\theta) = -\Delta H_m \ln\theta \quad (1 > \theta > 0)$$

(Heat of adsorption is dependent on surface coverage)

$$\theta \uparrow \;\Delta H \downarrow$$

(4) Freundlich (empirical)

$$\theta = a{p_a}^m$$

(For multi-component adsorption $\Rightarrow$ use Toth, Sips isotherms $\Rightarrow$ see the book by DO, 1998)

Different types of isotherms:

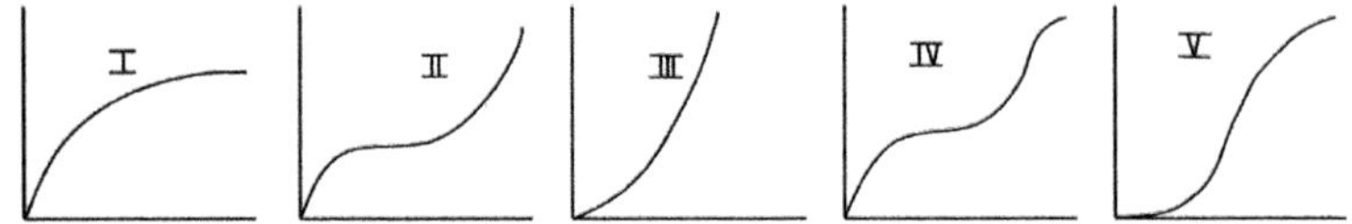

Physical adsorption (Physisorption)

- No bond formation
- Non-specific
- Forces between adsorbed and adsorbate molecules are weak (van der Waals forces)
- Heat of adsorption is small (0.5–5 kcal/mole)
- Activation energy is low
- T $\uparrow$ amount of adsorbed molecules $\downarrow$
- Coverage is multilayer
- Characterizes the surface

Chemical adsorption (Chemisorption)

- Bond formation
- Specific
- Forces are strong (Covalent, valence forces)
- Heat of adsorption is large (> 10 kcal/mole)
- Can be low or high
- T $\uparrow$ amount $\downarrow$ or $\uparrow$
- monolayer
- Basic component of catalytic effect

Fundamental Mechanism of Heterogeneous Reaction

Rate equations: Langmuir-Hinshelwood model/Eley-Rideal model (most common)

$$A + B \rightleftharpoons C + D$$

(Reaction only in the presence of catalyst; non-elementary reaction)

$$X\ (sites) \qquad : catalyst\ surface$$

Elementary steps:

(1) Adsorption
(2) Surface-reaction
(3) Desorption

$$[AX] \quad [BX] \rightarrow [CX] \quad [DX]$$

$L - H$ model assumes that the step (2) or surface reaction is the slowest or rate-determining step. Other possibilities also exist. See the book by Froment-Bischoff for details.

1.

$$\left.\begin{array}{c} A + X \underset{k_{-1}}{\overset{k_1}{\rightleftharpoons}} AX \\[1em] B + X \underset{k_{-2}}{\overset{k_2}{\rightleftharpoons}} BX \end{array}\right\} adsorption$$

$$AX + BX \underset{k_{-3}}{\overset{k_3}{\rightleftharpoons}} CX + DX : \quad \text{Surface reaction/re} - \text{arrangement}$$

(slowest; other steps are at equilibrium; they are fast enough to reach equilibrium and $r = 0$)

$$\left.\begin{array}{c} CX \underset{k_{-4}}{\overset{k_4}{\rightleftharpoons}} C + X \\[1em] DX \underset{k_{-5}}{\overset{k_5}{\rightleftharpoons}} D + X \end{array}\right\} desorption$$

(Note that there is "X" balance in all steps;

$$O_2 + 2X \rightleftharpoons 2OX \quad \text{or} \quad O_2 + X \rightleftharpoons O_2X$$

2.

$$r = k_3(AX)(BX) - k_{-3}(CX)(DX) \quad \text{(rate determining step)}$$

(Replace all AX, BX, ... etc., in terms of major species A, B, C, D)

3.

$$K_1 = \frac{(AX)}{(A)(X)}; \quad K_2 = \frac{(BX)}{(B)(X)}; \quad K_4 = \frac{(CX)}{(C)(X)}; \quad K_5 = \frac{(DX)}{(D)(X)}$$

(*Note*: K_4, K_5 are defined in reverse manner to be consistent with reactions 1 and 2).

4. Substitute:

$$r = k_{-3}\left[K_1 K_2 K_3(A)(B) - K_4 K_5(C)(D)\right] X^2$$

X is still unknown:

5.

$$X + AX + BX + CX + DX = X_o \; (\textit{property of a catalyst})$$

$$\uparrow \qquad \underbrace{}_{\textit{occupied}} \qquad (\textit{total})$$

$$\textit{free}$$

$$X = \frac{X_o}{1 + \sum_i K_i C_i}$$

$$= \frac{k_{-3} X_o^2 \left[K_1 K_2 K_3(A)(B) - K_4 K_5(C)(D) \right]}{\left(1 + \sum_i K_i C_i\right)^2}$$

Note: denominator term appears as the "inhibitor" of the rate. It appears to decrease the rate.

Physically, A and B are adsorbed on the sites, occupying some of the sites, and therefore, inhibiting the reaction. Adsorption is an essential part of the reaction, but at the same time it decreases the rate.

If equilibrium, $r = 0$

$$\frac{(C)(D)}{(A)(B)} = K_{eq} = K_3 \frac{K_1 K_2}{K_4 K_5}$$

$K_1 \ldots K_5$ are the equilibrium constants of the elementary steps; does not have physical or fundamental significance (lumped parameter)

(*Note*: Any elementary reaction, even if it is not at equilibrium has an equilibrium constant. Therefore, K_3 is well defined even if r in the step 2 is non-zero)

Note "r" can also be written as

$$r = \frac{K_f(A)(B) - K_b(C)(D)}{\left(1 + \sum K_i C_i\right)^2}$$

Suppose $\sum K_i C_i \ll 1$

$$r \approx K_f(A)(B) - K_b(C)(D)$$

It appears as an elementary reaction with K_f and K_b as the forward & reversible rate constants, respectively! But K_f and K_b are lumped parameters, and cannot be written as

$$K_f = K_{fo} \exp(-\Delta H/RT), \ etc.$$

These are the features of a catalytic reaction (non-elementary).

$A \rightarrow R$ (interphase transport is fast)

(Assume nth -order) or $C_s \rightarrow C_b$ ($k_m a \uparrow$) (the surface sees the bulk phase concentration)

Reaction occurs within the material as the reacting species "A" diffuses inside the pores.

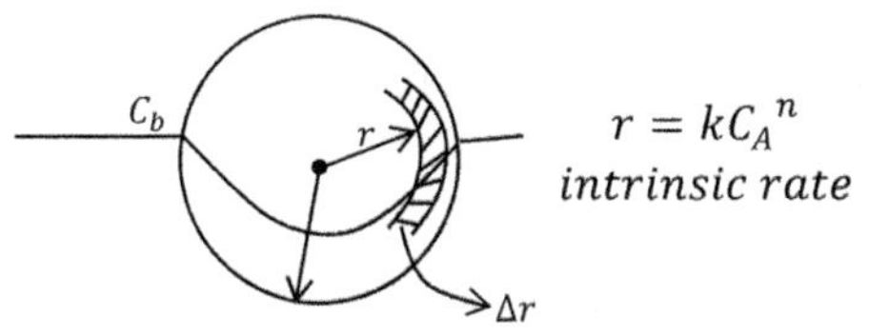

Isothermal Case

Take a CV $\left(4\pi r^2 \Delta r\right)$ between r and $r + \Delta r$:

Species balance:

$$\cancel{\frac{\partial C_A}{\partial t}}_{0} + \cancel{\mathbf{v}_r . \nabla C_A}_{0} \equiv D_{\text{eff}} \nabla^2 C_A - kC_A^n$$

$$D_{\text{eff}} \nabla^2 C_A = kC_A{}^n \quad \left(\frac{\text{moles}}{s-m^3}\right) \quad (0 \leq r \leq r_o)$$

$$\frac{D_{eff}}{r^2} \frac{d}{dr}\left(r^2 \frac{dC_A}{dr}\right) = kC_A{}^n$$

BC 1 $r = 0$, $\frac{dC_A}{dr} = 0$ (symmetric BC)

BC 2 $r = r_o$, $C_A = C_{A,b}$ $(C_A|_{r=r_o} = C_b)$

© The Author(s) 2025

N. Verma, *Chemical Reaction Engineering*,

https://doi.org/10.1007/978-3-031-88691-1_24

Non-dimensionalize

$$z = \frac{r}{L} \quad \& \quad f = \frac{C_A}{C_{A,b}}$$
$$L = \frac{1}{\underbrace{a}_{m^2/m^3}} = \frac{r_o}{3} \ \text{(sphere)}$$

substitute,

$$\frac{D_{eff} C_{A,b}}{L^2} \nabla^2 f = k f^n C_{A,b}{}^n$$

$$\nabla^2 f = \frac{L^2 k}{D_e} C_{A,b}{}^{n-1} f^n$$

$$\nabla^2 f = \phi^2 f^n; \quad \phi \equiv \text{Thiele modulus} = L\sqrt{\frac{k C_{A,b}{}^{n-1}}{D_e}}$$

BCs

$z = 0, \quad \nabla f = 0 \quad \text{(symmetry)}$

$z = 3, \quad f = 1 \quad \text{(surface)}$

1.

$$\phi \equiv \frac{\text{kinetic rate}}{\underbrace{\text{mass transfer rate}}_{\text{intraphase (pore-diffusion)}}} = L\sqrt{\frac{k C_{A,b}^{n-1}}{D_e}}$$

2. Define,

$$\eta = \text{effectiveness factor} = \frac{R_{\text{obs/global/apparent}}}{R_b} \le 1 \quad \text{for} +\text{ve order}$$

where, R_b = rate without diffusion (max at the surface)

$$R_{obs} = \eta \left(k C_{A,b}^n \right) \Rightarrow \eta_{\text{intra}} = \eta(\phi, n)$$

compare to $\eta_{\text{inter}} = \eta \left(D_a, n \right)$

- First solve the equation (numerically) $\Rightarrow f(z)$ or $C_A(r)$
- Calculate "R_{obs}"

Alternatively, calculate ϕ, η_{intra} and then R_{obs}.

Special Case: First-order reaction (analytical solution is available)

$$D_e \nabla^2 f = kf$$
$$D_e \left(\frac{d^2 f}{dz^2} + \frac{2}{z} \frac{df}{dz} \right) = kL^2 f$$
$$\frac{d^2 f}{dz^2} + \frac{2}{z} \frac{df}{dz} = \phi^2 f; \quad \phi = L\sqrt{\frac{k}{D_e}} \quad \left(L = \frac{r_o}{3} \text{ for sphere} \right)$$

BCs $z = 0, \quad \nabla f = 0$

$\quad\quad z = 3, \quad f = 1$

$$(z = r/L = r_o/(r_o/3) = 3)$$

Let $g = fz$; $f = g/z$. Differentiate f twice with respect to z, simplify and substitute in the equation above to show the following transformation:

$$\frac{d^2 g}{dz^2} - \phi^2 g = 0$$

BCs $z = 0, \quad g = 0$

$\quad\quad z = 3, \quad g = 3$

$$g = C_1 \sinh \phi z + C_2 \cosh \phi z$$

$$C_2 = 0, \quad C_1 = 3/\sinh 3\phi \quad \text{(apply BCs)}$$
$$g = \frac{3 \sinh(\phi z)}{\sinh(3\phi)}$$
$$f = \frac{3 \sinh(\phi z)}{z \sinh(3\phi)} : \text{(dimensionless solution)}$$

$$\frac{C_A}{C_{A,b}} = \frac{3 \sinh(\phi r/L)}{r/L \sinh(3\phi)} = \frac{r_0 \sinh(\phi 3r/r_0)}{r \sinh(3\phi)}$$

Thus, $C_A(r)$ is determined.

Calculate the reaction rate by equating the rate of species arriving at the surface of the catalyst to that of species consumed/reacted in the catalyst volume under SS conditions.

$$R \left(\frac{\text{moles}}{s} \right) = \int_0^{r_o} \underbrace{k C_A(r)}_{\text{moles/s-m}^3} \underbrace{4\pi r^2}_{m^3} dr$$

$$= -D_e \frac{dC_A}{dr} \bigg|_{r=r_o} 4\pi r_o^2$$

("A" is consumed in the catalyst pores and flux of A arrives at the surface).

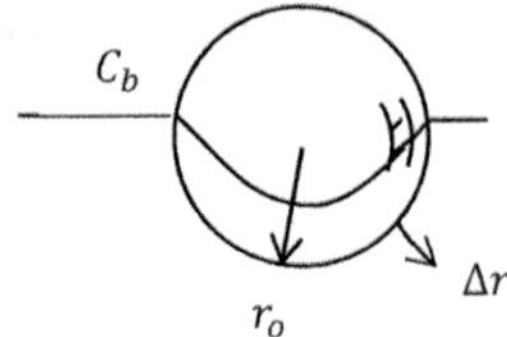

Note that one can also write:

$$R_{\text{global}} = k\overline{C_A(r)}$$

based on the volume average concentration inside the catalyst.

It is easier to differentiate than to integrate (use the second expression)

$$\eta_{\text{intra}} = \frac{R_{\text{global}}}{R_b} = \frac{-D_e \frac{dC_A}{dr}\Big|_{r=r_o}\, 4\pi r_o{}^2}{\left(kC_{A,b}\right)\frac{4}{3}\pi r_o{}^3}$$

$$= D_e \frac{d\left(C_A/C_{A,b}\right)}{d(r/L)}\Bigg|_{r=r_o} \times \frac{1}{L^2 k} \quad (L = r_o/3)$$

$$= \frac{1}{\phi^2}\frac{df}{dz}\Bigg|_{z=3}$$

Already we have the solution $f(z)$ from the previous exercise; substitute to obtain:

$$\eta_{\text{intra}} = \frac{1}{\phi}\left[\frac{1}{\tanh 3\phi} - \frac{1}{3\phi}\right]$$

$$\phi = L\sqrt{\frac{k}{D_e}} \qquad \begin{array}{c} \searrow\, 1/s\ (1^{st}\,order) \\ \downarrow \\ m \end{array} \quad \searrow\, m^2/s$$

Once we know η_{intra} calculate

$$R_{\text{global}} = \eta_{\text{intra}} \times \left(kC_{A,b}\right)\ \text{mole/s-m}^3$$

Plot:

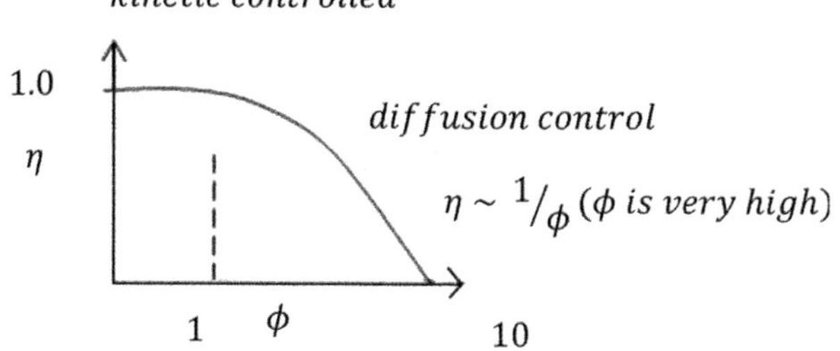

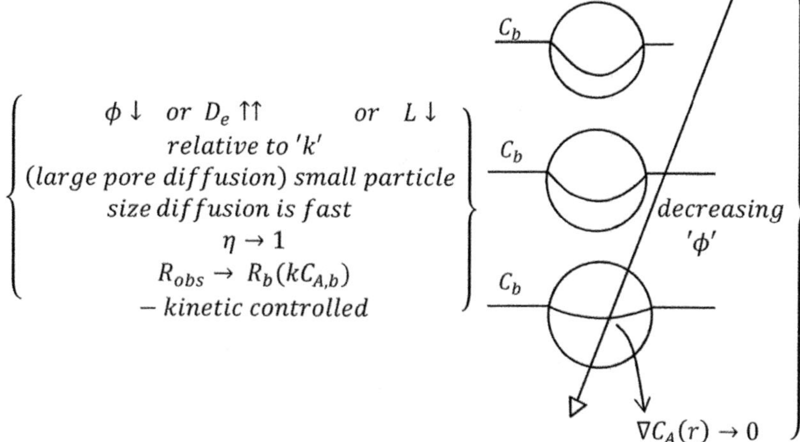

Alternatively,

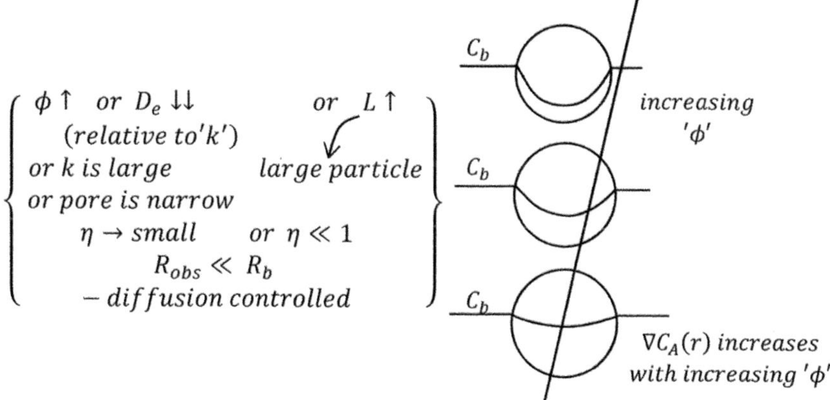

Note: Graphical solutions are available for different shapes of the materials and different orders of reaction.

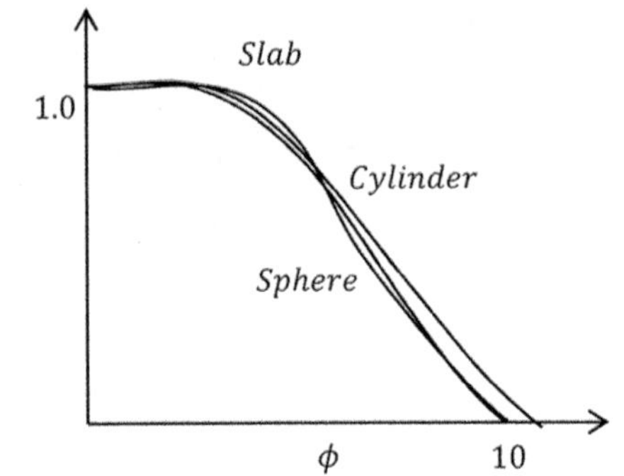

and

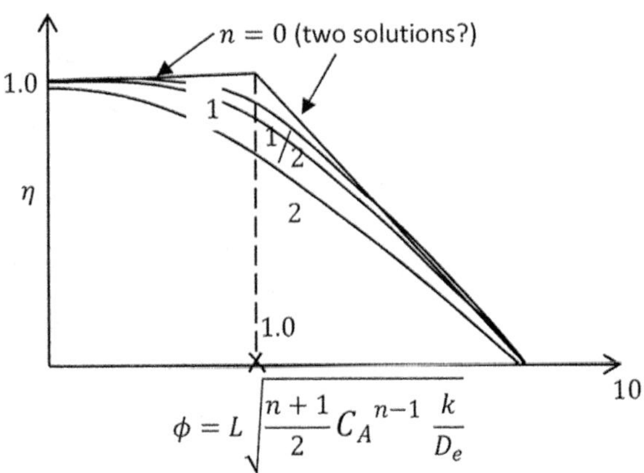

$$\phi = L\sqrt{\frac{n+1}{2}C_A{}^{n-1}\frac{k}{D_e}}$$

Again, calculate ϕ first and then η_{intra} calculate

$$R_{\text{global}} = \eta_{\text{intra}} \times \left(C_{A,b}{}^{n}\right) \text{ mole/s-m}^3$$

We will close the discussion on intraphase transport + kinetics in this lecture.

(Any mistakes noted in the book, or comments may be communicated to vermanishith@gmail.com)